MODULAR SERIES ON SOLID STATE DEVICES

Gerold W. Neudeck and Robert F. Pierret, Editors

VOLUME II

The PN Junction Diode

GERALD W. NEUDECK

Purdue University

ADDISON-WESLEY PUBLISHING COMPANY

READING, MASSACHUSETTS
MENLO PARK, CALIFORNIA
LONDON • AMSTERDAM
DON MILLS, ONTARIO
SYDNEY

This book is in the
Addison-Wesley Modular Series on Solid State Devices

Library of Congress Cataloging in Publication Data

Neudeck, Gerold W.
The PN junction diode.

(Modular series on solid state devices ; v. 2)
Bibliography: p.
Includes index.
1. Diodes, Semiconductor. I. Title. II. Title: P. N. junction diode. III. Series: Pierret, Robert F. Modular series on solid state devices ; v. 2.
TK 7871.86.N48 621.3815'22 81-14979
ISBN 0-201-05321-7 AACR2

Reprinted with corrections, July 1983

ISBN 0-201-05321-7
BCDEFGHIJ-AL-89876543

Foreword

Solid state devices have attained a level of sophistication and economic importance far beyond the highest expectations of their inventors. The bipolar and field-effect transistors have virtually made possible the computer industry which in turn has created a completely new consumer market. By continually offering better performing devices at lower cost per unit, the electronics industry has penetrated markets never before addressed. One fundamental reason for such phenomenal growth is the enhanced understanding of basic solid state device physics by the modern electronics designer. Future trends in electronic systems indicate that the digital system, circuit, and IC layout design functions are being merged into one. To meet such present and future needs we have written this series of books aimed at a qualitative and quantitative understanding of the most important solid state devices.

Volumes I through IV are written for a junior, senior, or possibly first-year graduate student who has had a reasonably good background in electric field theory. With some deletions these volumes have been used in a one semester, three credit-hour, junior-senior level course in electrical engineering at Purdue University. Following this course are two integrated-circuit-design and two IC laboratory courses. Each volume is written to be covered in 12 to 15 fifty-minute lectures.

The individual volumes make the series useful for adoption in standard and non-standard format courses, such as minicourses, television, short courses, and adult continuing education. Each volume is relatively independent of the others, with certain necessary formulas repeated and referenced between volumes. This flexibility enables one to use the series continuously or in selected parts, either as a complete course or as an introduction to other subjects. We also hope that students, practicing engineers, and scientists will find the volumes useful for individual instruction, whether it is for reference, review, or home study.

A number of the standard texts on devices have been written like encyclopedias, packed with information, with little thought to how the student learns or reasons. Texts that are encyclopedic in nature are difficult for students to read and are often barriers to their understanding. By breaking the material into smaller units of information and by writing for students (rather than for our colleagues) we hope to enhance their understanding of solid state devices. A secondary pedagogical strategy is to strike a healthy balance between the device physics and practical device information.

The problems at the end of each chapter are important to understanding the concepts presented. Many problems are extensions of the theory or are designed to reinforce particularly important topics. Some numerical problems are included to give the reader an intuitive feel for the size of typical parameters. Then, when approximations are stated or assumed, the student will have confidence that certain quantities are indeed orders of magnitude smaller than others. The problems have a range of difficulty, from very simple to quite challenging. We have also included discussion questions so that the reader is forced into qualitative as well as quantitative analyses of device physics.

Problems, along with answers, at the end of the first three volumes represent typical test questions and are meant to be used as review and self-testing. Many of these are discussion, sketch, or "explain why" types of questions where the student is expected to relate concepts and synthesize ideas.

We feel that these volumes present the basic device physics necessary for understanding many of the important solid state devices in present use. In addition, the basic device concepts will assist the reader in learning about the many exotic structures presently in research laboratories that will likely become commonplace in the future.

West Lafayette, Indiana

G. W. Neudeck
R. F. Pierret

Contents

3 The Diode Volt–Ampere Characteristic

4 Deviations from the Ideal Diode

5 *p-n* Junction Admittance

Introduction

The p-n junction diode is the most fundamental of all the semiconductor devices, and for this reason we devote an entire volume to it. So basic is the theory of operation that many engineers have stated, "to understand the p-n junction qualitatively and quantitatively allows you to understand the majority of all solid state devices." This common thread of application from device to device builds confidence and competence as you progress into the more exotic device structures. For example, the bipolar transistor is two very closely spaced p-n junctions; the solar cell is a special p-n junction made to absorb sunlight; the silicon-controlled rectifier (SCR) and triacs are multiple p-n junctions; the light emitting diode is a p-n junction designed to emit light efficiently; the MOS and junction field effect transistors also contain p-n junctions. The list could be continued, but the main point is that a thorough study of the p-n junction is a direct path to the other solid state devices including all of the present work in integrated circuits. Most of the concepts and many of the equations developed for the diode will apply in a very direct and straightforward way to these other deivces.

The goal of this volume is to build a firm foundation in p-n-junction theory from a conceptual and mathematical viewpoint. This will allow the reader not only to appreciate the diode but—and this is even more important—also to have the necessary background for understanding other present and future devices.

This volume begins by discussing how p-n junctions are fabricated in single crystalline semiconductors. The wafer processing steps used in the fabrication of the diode are also basic to the production of other solid state devices such as the transistor and integrated circuit. Chapter 2 details the internal workings of the junction, in particular the depletion region. The case of thermal equilibrium is contrasted with the cases of forward and reverse bias. Once all the electrons and holes have been accounted for, the charge density, electric field, and potential are developed. The case of forward and reverse bias for the ideal diode and the accompanying volt–ampere (V–I) characteristic are examined in Chapter 3. Chapter 4 is devoted to deviations from the ideal diode: avalanche breakdown, generation–recombination, and high-level injection. Chapter 5 discusses the small signal admittance, conductance, junction capacitance, and diffusion capacitance. Chapter 6 develops the large signal switching transient for turning the diode "off" abruptly when forward biased and the turn-on transient when reverse biased.

1 / Introduction to the Diode

The p-n junction diode is produced by forming a single crystal semiconductor such that part of the crystal is doped p-type and the other part is doped n-type. Junctions are classified by how the transition from the p-type to n-type is developed within the single crystal. The junction is said to be *abrupt* when the transition is extremely narrow. The *graded* junction is a junction where the transition region is "spread out" over a larger distance. In later chapters we will indicate how the nature of the junction's "abruptness" affects the electrical characteristics.

1.1 *p-n* JUNCTION FABRICATION

Abrupt p-n junctions are formed by alloying and epitaxial growth. Graded junctions can be produced by gaseous diffusion of impurities or by ion implantation.

1.1.1 Alloyed Junctions

The most ideally abrupt of the p-n junctions results from a fabrication process called *alloying*. In the alloying process a p^+-n junction is formed by starting with an n-type semiconductor wafer, placing on the surface of the wafer a doping impurity (often a metal), and heating until the impurity reacts with the semiconductor. Figure 1.1 illustrates this process for n-type silicon and the doping impurity of aluminum (Al). Remember that aluminum acts as an acceptor in silicon. When heated to an alloying temperature of 580°C the Al softens and "eats up" some of the silicon below the molten Al; that is, it forms a solution of Al and Si atoms. When the temperature is carefully lowered, the silicon begins to regrow on the atomic sites of the host n-type silicon wafer. However, the large number of Al atoms present in the Al–Si solution are far greater than the n-type impurities and the regrowth region is changed to a heavily doped p-type region. In fact, it is degenerate p-type, or p^+. The maximum number of Al atoms possible in the

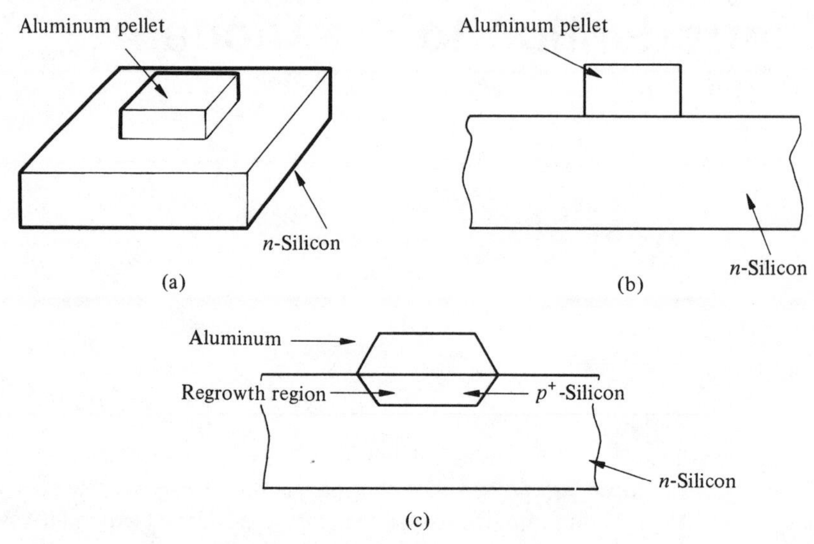

Fig. 1.1 (a) Alloying Al to *n*-silicon; (b) cross section of (a); (c) after alloying.

silicon, under normal conditions, is called its *solid solubility limit* ($\simeq 3 \times 10^{18}/\text{cm}^3$ Al in Si at 580°C). The net result of the alloying process is a p^+-n junction with a very sharp (abrupt) transition. As illustrated by Fig. 1.1 the remainder of Al, that which was not incorporated into the silicon, is left at the surface and is a convenient electrical contact to the p^+ region. Also note that the Al has selectively formed the surface geometry (area) of the p^+-n junction.

1.1.2 Epitaxial Growth

Epitaxial growth of a semiconductor layer on top of a single crystal semiconductor substrate is another method by which an abrupt junction can be formed. Epitaxial growth is accomplished by heating the host wafer, say *n*-type silicon, and passing a gas containing silicon tetrachloride ($SiCl_4$) and hydrogen (H) over the surface at a controlled flow rate. The gases react and deposit silicon atoms on the surface of the host wafer. Because the temperature is usually in excess of 1000°C, the deposited Si atoms have sufficient energy and mobility to align themselves properly to the host-wafer crystal lattice. This forms a continuation of the lattice up from the original surface. Typical epitaxial-layer growth rates are $\simeq 1$ micron per minute.

Impurity atoms, in the form of a gaseous compound, can be added to the carrier gas during epitaxial growth to form either *n*-type or *p*-type layers. Starting with an *n*-type host wafer and growing a *p*-type epitaxial (epi) layer forms a fairly abrupt *p*-*n* junction.

Other combinations are, of course, possible, such as growing an n-type epitaxial layer on a p-type substrate.

The epitaxial process is used extensively in making integrated circuits (IC's). The p-n diode formed in the "epi" process is kept reverse biased and provides for circuit isolation from the substrate (host wafer). More recently, epitaxy has been used in the formation of SOS structures, where SOS stands for silicon–on–sapphire or silicon–on–spinel. Spinels are various mixtures of MgO (magnesium oxide) and Al_2O_3 (aluminum oxide) and are closely related to sapphire. To make a long story short, doped silicon is epitaxially grown on substrates of sapphire or spinel. The incentive for this procedure is the exceptional insulator quality of the sapphire and spinel substrates in isolating circuits in IC designs requiring high-speed devices, especially large-scale integrated circuits (LSI).

1.1.3 Thermal Diffusion

Junctions in which the transition from p- to n-type silicon occurs over many atomic spacings are called *graded junctions*. An extremely important method, instrumental in the growth of the semiconductor industry, is that of gaseous (thermal) diffusion of the impurities directly into the host crystal. The impurities are introduced into an inert carrier gas and passed over the surface of the silicon wafer. Due to the high temperature and the large number of impurities at the surface, the impurity atoms migrate (diffuse) into the crystal. As one might expect the impurity distribution (concentration) is largest near the

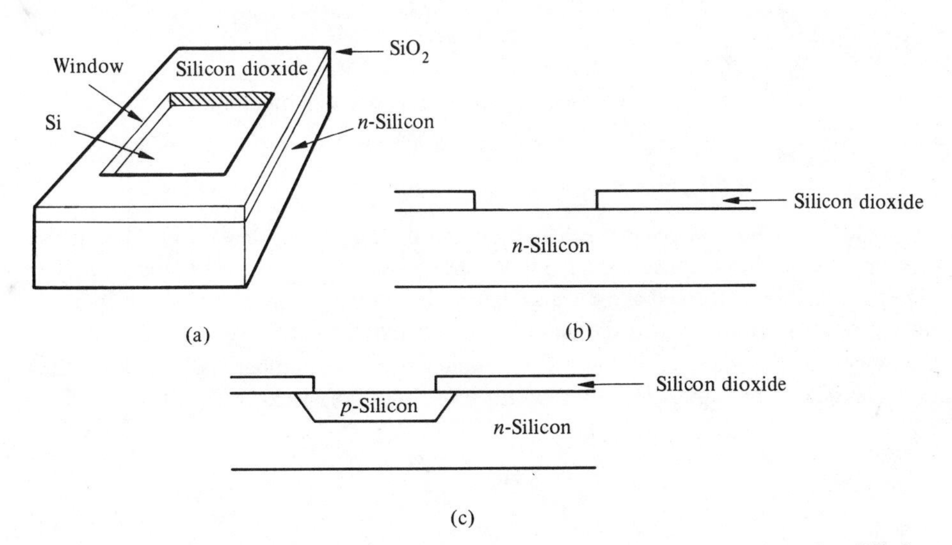

Fig. 1.2 (a) Window in SiO; (b) cross section before diffusion; (c) after p-impurity diffusion.

surface and is progressively smaller further into the crystal. The impurities are typically distributed in a complementary error function or as a Gaussian function (more on this later).

An important property of silicon is the ability of its natural oxide, silicon dioxide (SiO_2), to form in an oxidizing atmosphere. SiO_2 is a glass, and hence it is very impervious to moisture and other contaminants. It also serves as a barrier to the diffusion of the desired impurities and therefore allows precise geometric control of the p-n junction area. Figure 1.2 illustrates how the oxide is used to define the surface geometry of a diffused (graded) junction diode.

So important are the properties of Si–SiO_2 that without them the entire computer industry would not even exist as we know it today. Silicon dioxide has allowed us to build integrated circuits in large volume inexpensively; without it our systems would be limited in reliability, cost, and complexity. More will be said on this topic as each device is discussed in later chapters and volumes.

During diffusion, the concentration of impurities at the surface imposes a boundary condition on the diffusion process. If the *surface concentration* is kept constant at N_0 impurities/cm^3, then the *impurity distribution* is a complementary error function, as pictured in Fig. 1.3(a). Starting from the surface the impurity concentration decreases going into the semiconductor. The complementary error function is just another of those unusual functions whose value is often obtained from plots or tables. Equation (1.1) lists the mathematical form of the function.

$$N(x,t) = N_0 \operatorname{erfc}\left[\frac{x}{2\sqrt{Dt}}\right], \qquad (\#/\text{cm}^3) \qquad (1.1)$$

where

D is the diffusion constant and is a function of temperature,

t is the length of time the diffusion takes place,

N_0 is the surface concentration (#/cm^3).

It should be obvious that when the diffused impurities are greater than the host impurities, "compensation" has occurred and the host semiconductor is converted to the other type. The p-n junction is formed at a distance below the surface where $N_A = N_D$, as indicated in Fig. 1.3(b). This is called the *metallurgical junction* (x_j).

When the total number of impurities is limited to a fixed number, then the distribution is a Gaussian function as described by Eq. (1.2) and illustrated in Fig. 1.4.

$$N(x,t) = \frac{Q}{A\sqrt{\pi Dt}}\, e^{-x^2/(4Dt)}, \qquad (\#/\text{cm}^3) \qquad (1.2)$$

where

Q is the total number of impurities initially deposited near the surface,

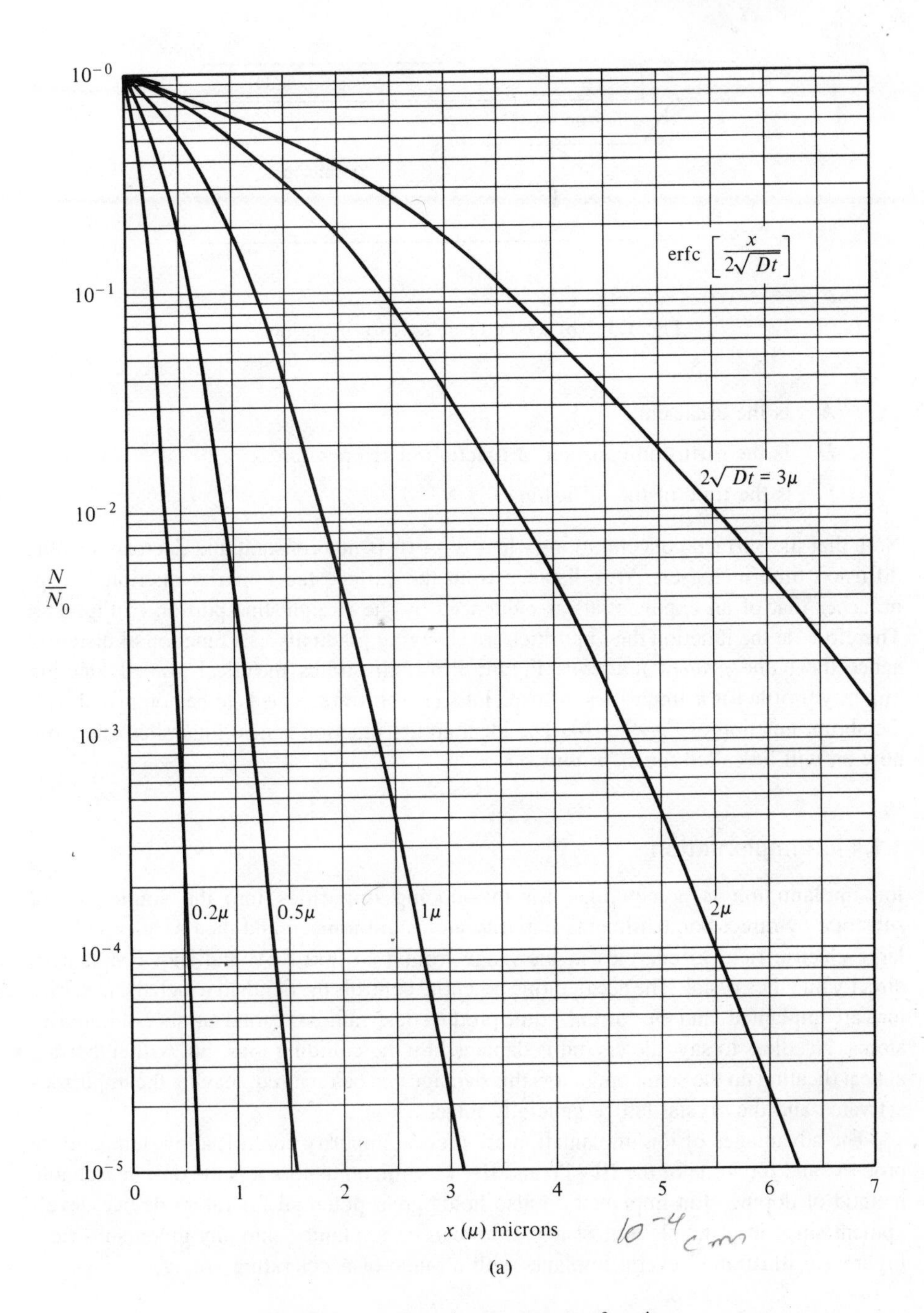

Fig. 1.3 (a) Complementary error function.

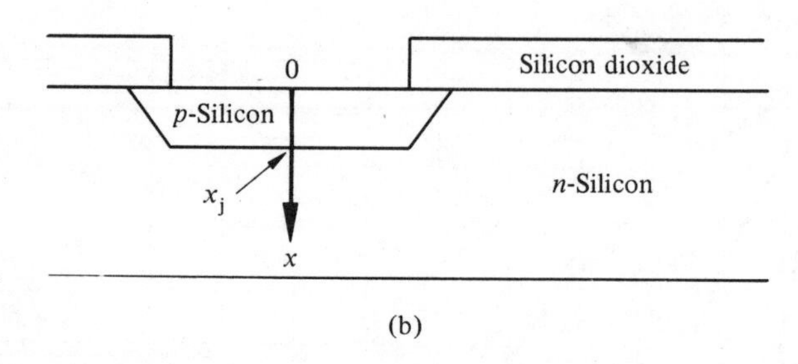

(b)

Fig. 1.3 (b) x-coordinate axis for impurities.

A is the area, cm^2,

D is the diffusion constant, a function of temperature,

t is the time of the diffusion.

Note that the surface concentration (where $x = 0$) is not constant and decreases as the diffusion time increases. At a distance from the surface the impurity distribution approaches that of an exponential, as evidenced by the straight-line portion of Fig. 1.4. Therefore, at the junction the impurities are changing gradually as a function of distance; hence the name *graded junction*. Figure 1.5(a) illustrates the ideal graded junction impurity profile for n impurities diffused into a p substrate. This is to be contrasted with the abrupt junction of Fig. 1.5(b). The ideal abrupt junction is also illustrated and from now on will be called the *step junction*.

1.1.4 Ion Implantation

Ion implantation is a technique for introducing impurities into the single crystal substrate by direct bombardment. Impurity atoms are ionized and then accelerated in a large electric field to energies in the range from 1 to 300 KeV and shot (implanted) directly into the crystal. The accelerating potential controls the depth to which the impurity ions are implanted, and the current–time product determines the total number of impurity atoms. Needless to say, the crystal is damaged by the colliding ions, but with a thermal anneal (heating up the semiconductor) this damage can be repaired, leaving the impurities activated and the crystal lattice generally intact.

The advantages of ion implantation are precise impurity control, a low-temperature process, and for some of the III–IV and II–VI semiconductors it is the only reasonable method of doping. Ion implantation also holds great potential for future device development since in principle almost any atom can be implanted into any given substrate. Figure 1.6 illustrates several implants with a range of accelerating voltages.

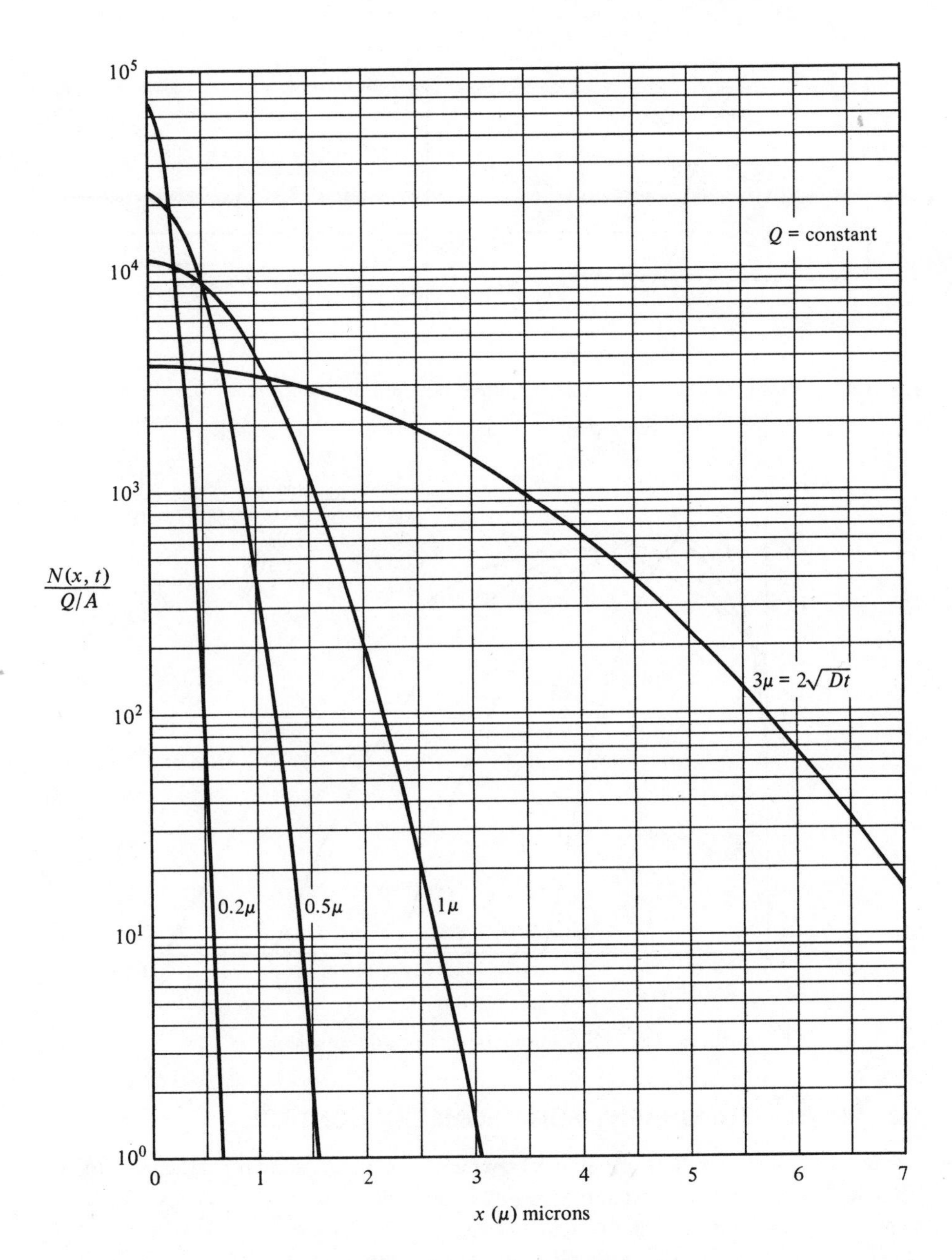

Fig. 1.4 Gaussian distribution of impurities.

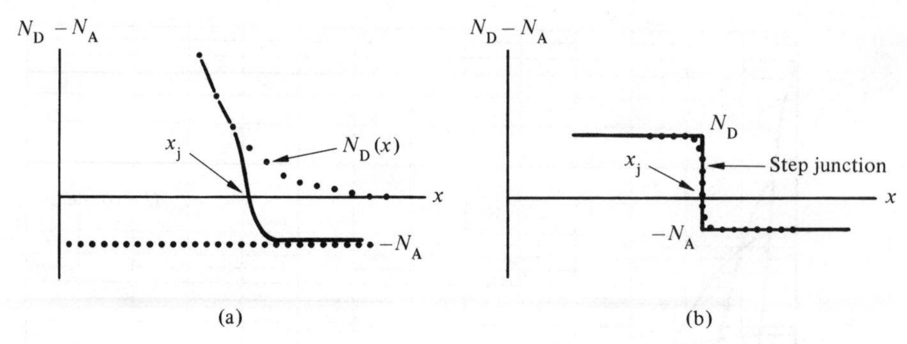

Fig. 1.5 Impurity profiles: (a) diffused n-impurities into uniform p-substrate; (b) abrupt n-p junction . . . . . , step junction ________.

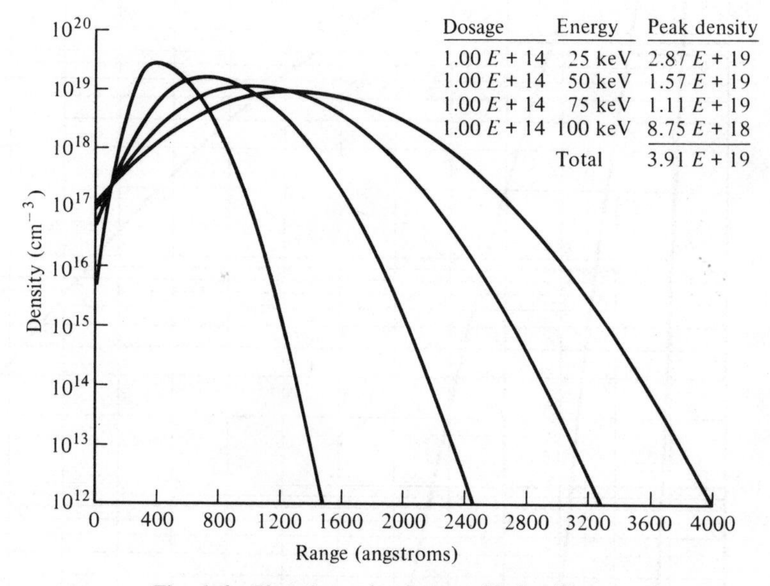

Fig. 1.6 Phosphorus-implant profile in silicon.

1.2 PHOTOLITHOGRAPHY FOR GEOMETRY CONTROL

The previous section indicated methods by which the surface geometry of the p-n junction could be controlled. The most universal of the methods is that of using a mask which acts as a barrier to the impurity atoms while openings or *windows* are made where impurity penetration is desired. For example, the thermal diffusion of impurities into silicon through a window in the SiO_2 is illustrated by Fig. 1.2. One might ask the question, "How can one make a window in the SiO_2 one-thousandth of an inch on a side?" The answer is by using a process called *photolithography*.

Photolithography makes use of an old idea which was called "photoengraving" and was developed to make very small machine parts. With the scale greatly reduced and aimed at engraving windows for IC manufacturing, the process has become quite sophisticated. Figure 1.7 illustrates the steps used in producing a window in SiO_2 and is used only as an example of the many possible approaches in present use.

A silicon wafer is placed in an oxidation furnace at about 1000 to 1200°C in a steam and/or oxygen atmosphere to grow a few thousand angstroms ($\approx$5000 Å) of silicon dioxide (SiO_2) on its surface. The gases furnish the oxygen component and the silicon wafer furnishes the Si atoms. This results in the SiO_2 using up some of the silicon wafer surface. When finished, the once metallic gray surface of the silicon will have a range of colors depending on the thickness of the silicon dioxide. This is the reason pictures of IC's are multicolored with regions of blue, green, pink, etc.

The SiO_2 is next coated with a liquid photoresist material that is impervious to acid etching, but is sensitive to certain types of light, usually ultraviolet light. The liquid photoresist is typically applied by spinning the wafer at high speeds in order to produce a uniformly thin layer. After baking to harden, the photoresist is ready for exposure. The hardened photoresist is similar to a photographic emulsion. Figure 1.7(a) illustrates the next step, that of placing a photomask (similar to a developed picture negative) over the wafer and exposing with ultraviolet light the photoresist material in those areas where the SiO_2 is to remain. The light causes those areas of photoresist to polymerize (harden further). The photoresist is then developed much as you would an ordinary photographic

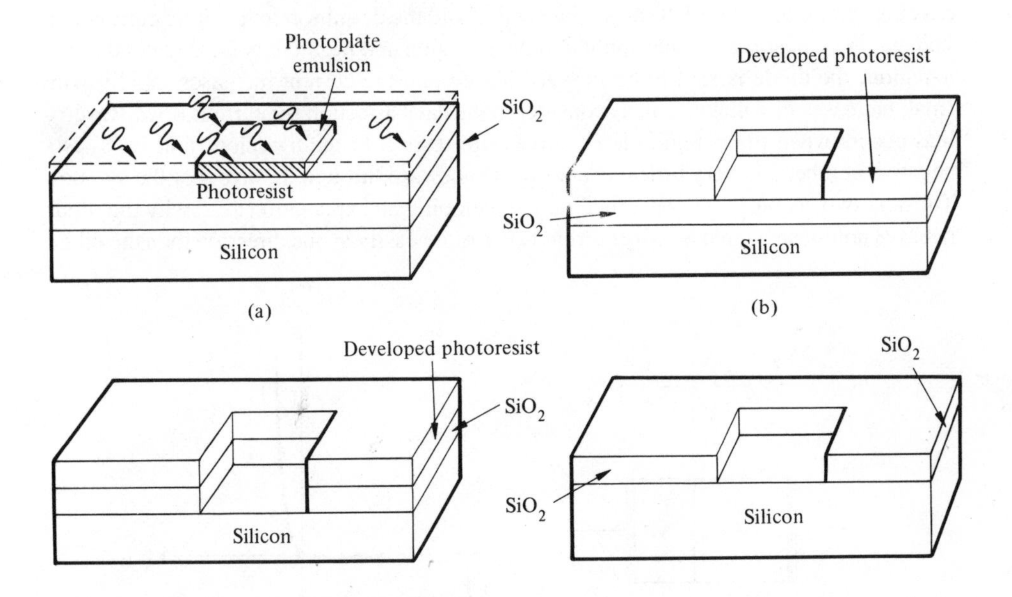

Fig. 1.7 Photolithography: (a) exposure of photoresist; (b) developed photoresist; (c) etched silicon dioxide; (d) window in the silicon dioxide.

film. The area not exposed to ultraviolet light is washed away, leaving the SiO_2 bare only in the unexposed areas of Fig. 1.7(a), as illustrated by Fig. 1.7(b).

A hydrofluoric-based acid that does not attack the photoresist or pure silicon is used to etch the window in the SiO_2 as shown in Fig. 1.7(c). The photoresist is then stripped from the SiO_2, leaving a window in the SiO_2 open to the silicon surface. The wafer is now ready for the thermal diffusion of impurity atoms through the window in the SiO_2, as illustrated in Fig. 1.7(d).

The basic process of photolithography is applied in many different combinations and we have presented only one. For example, the photoresist could be of the opposite type where, with light exposure, it becomes soluble in the developer. Hence, those areas exposed to the ultraviolet light would be removed. Regardless of the specific details, a photomask, photoresist, light, and acid etch are used to open windows in specific areas of the SiO_2.

In the new era of large-scale integrated circuits (LSI), extending into very large-scale LSI, called VLSI, the wavelength of light is too long for the geometries needed. Fringing becomes a problem, and x-ray or electron beam photolithography is used to expose special photoresists. For ion implantation, photolithography is used to open windows in the SiO_2, photoresist, or metals that are used to mask the surface of the silicon.

1.3 DIODE SYMBOLS AND DEFINITIONS

The standard *p*-*n* junction diode has as its symbol an arrow to indicate the direction of easy current flow. Figure 1.8 shows the symbol and the definition for positive current and voltage. With positive voltage applied to the *p*-region and negative voltage applied to the *n*-region, the diode is said to be forward biased and the current increases rapidly with small increases in voltage. This is considered the easy direction for current flow. Reverse bias occurs when the *p*-region is negative with respect to the *n*-region, that is, V_A is a negative number and very little current flows backward through the arrow of the symbol. The next two chapters are concerned with developing an explanation as to why the diode behaves nonlinearly; that is, why current flows more easily in one direction than the other.

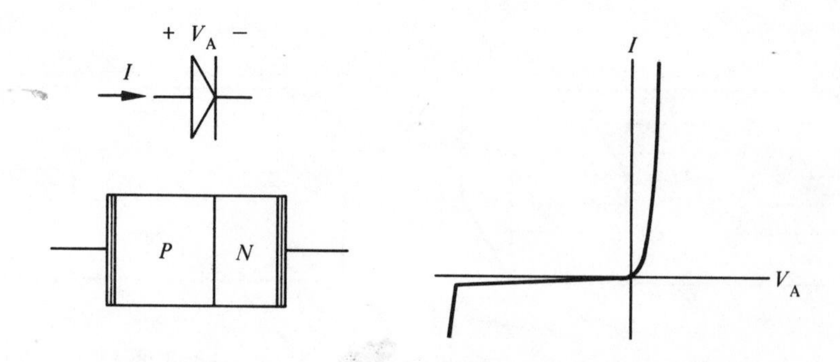

Fig. 1.8 The *p*-*n* junction diode polarity.

PROBLEMS

To evaluate the complementary error and Gaussian function, use Fig. P1.0.

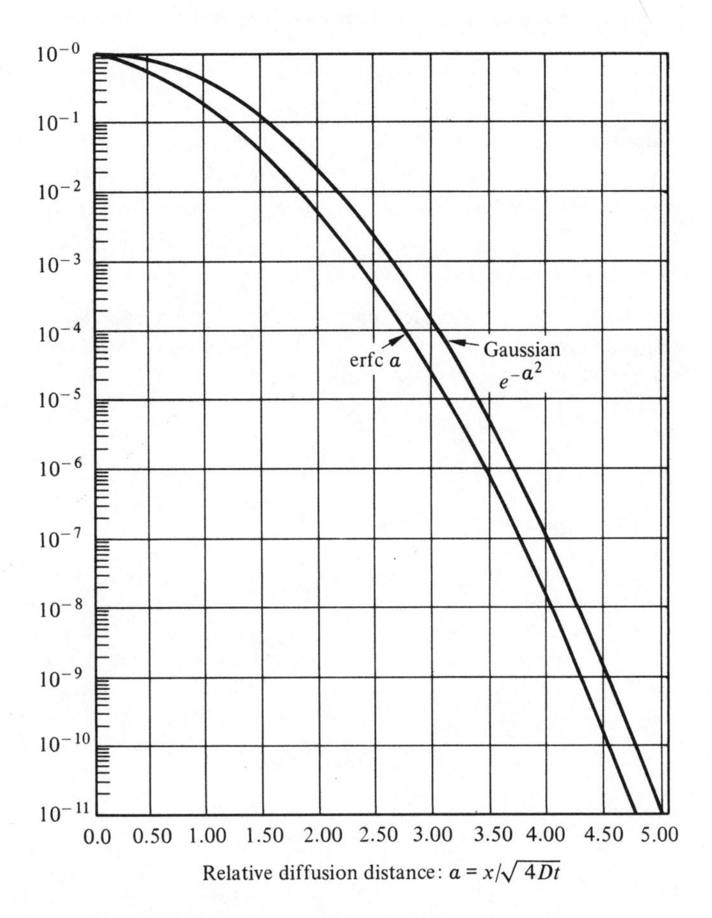

Fig. P1.0 Convenient normalized curves for erfc a and e^{-a^2}. (After D. J. Hamilton and W. G. Howard, *Basic Integrated Circuit Engineering,* McGraw-Hill Book Co., New York, 1975.)

1.1 If the surface concentration is constant at $10^{20}/\text{cm}^3$ boron atoms when diffusing into a uniformly doped n-substrate of $10^{16}/\text{cm}^3$ and the diffusion constant for boron at 1050°C is $10^{-13}\text{cm}^2/\text{sec}$:

(a) calculate the junction depth if the diffusion continues for 1 hr and 44 min.;

(b) if the diffusion time is four times as long, what is x_j?

(c) if the background doping were $10^{17}/\text{cm}^3$, repeat parts (a) and (b).

1.2 If boron were diffused from a limited source, $Q/A = 10^{16}/\text{cm}^2$, into a uniformly doped substrate of $N_D = 10^{16}/\text{cm}^3$ for 1 hr and 44 min:

(a) calculate the surface concentration;

(b) calculate the junction depth;

(c) calculate x_j after increasing the time fourfold.

1.3 Sketch the cross sections for photolithography similar to Fig. 1.7, given a photoresist that is of the opposite type to Fig. 1.7.

1.4 Uniformly doped Si, $N_D = 10^{15}/\text{cm}^3$, is thermally diffused with a p-type impurity at a temperature where $D = 10^{-13}\text{cm}^2/\text{sec}$. If from a limited source of impurities of $Q/A = 10^{14}/\text{cm}^2$:

(a) calculate the time necessary for $x_j = 3 \times 10^{-4}\text{cm}$. What is the surface concentration at the end of this time?

(b) Repeat part (a) for $x_j = 6$ microns.

1.5 To check the validity of the linear-graded junction of Fig. 1.5(a), plot part (a) of Problem 1.4 from 2.8μ to 3.2μ and compare it to a *straight line*. Use a least-squares fit to a straight line to find a correlation coefficient. Plot every $0.05 \times 10^{-4}\text{cm}$.

2 / Junction Statics

In this chapter we focus our attention on the transition region between the p and n regions of the diode under static biasing conditions, examining, in turn, thermal equilibrium ($V_A = 0$), forward bias ($V_A > 0$), and reverse bias ($V_A < 0$). The region near the metallurgical junction, the transition region, is often called the *depletion region* since, as is discussed in Section 2.1, the mobile carriers in the region are reduced in number, that is, depleted in population compared to the *bulk regions* far from the junction. We first consider the junction under equilibrium conditions qualitatively to establish the source of the charge density, electric field, and potential in the depletion region. With these quantities established, the concept of the *built-in potential* (V_{bi}) is introduced and treated quantitatively. The *depletion approximation* is next invoked to solve Poisson's equation for $\mathscr{E}(x)$ and $V(x)$, assuming a step junction doping profile and $V_A = 0$. The step junction analysis is subsequently extended to forward and reverse bias. The last section outlines the analysis for the linearly graded junction and presents the results for the electric field and potential.

2.1 QUALITATIVE EQUILIBRIUM ELECTROSTATICS

Before discussing the static properties of the junction, let us clearly specify the device configuration assumed throughout most of the chapter. Figure 2.1(a) illustrates the as-fabricated abrupt p-n junction; part (b) of the figure is a one-dimensional approximation to what is clearly a three-dimensional device. The one-dimensional approximation simplifies the "equations of state" to one spatial variable and facilitates closed-form solutions of the differential equations. Such an assumption is justified if the major variables change rapidly in only one direction. The entire list of major assumptions includes

1. a one-dimensional device;
2. a metallurgical junction at $x = 0$ (see Fig. 2.1(b));

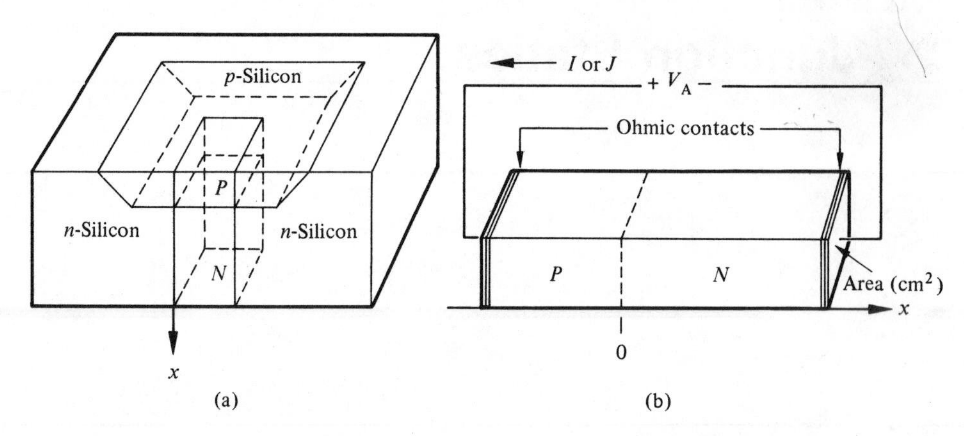

Fig. 2.1 (a) Abrupt junction *p-n* diode; (b) one-dimensional step junction.

3. a step junction from N_A to N_D with uniformly doped *p* and *n* regions (see Fig. 2.2(a));
4. perfect ohmic contacts far removed from the metallurgical junction.

The discussions in Volume I were principally concerned with homogeneously (uniformly) doped semiconductors. In the step junction diode, sections of the semiconductor are uniformly doped, but the doping is different on the opposite sides of the metallurgical junction. Our initial goal will be to determine and quantify the effects of this discontinuity in doping. We will begin with a qualitative examination of the structure under equilibrium conditions.

In our particular case "equilibrium" means no applied voltage ($V_A = 0$), no light shining on the device, no thermal gradients (uniform temperature), and no applied magnetic or electric fields. The situation external to the device is clearly of little interest. The internal situation, on the other hand, is quite interesting. In an attempt to ascertain the internal situation, let us first assume that *charge neutrality* exists everywhere in the device. If this were the case for the step junction of Fig. 2.2(a), the mobile carrier concentrations would be as pictured in Fig. 2.2(b) and (c). However, note that the holes on the *p*-side (p_p)* may be, for example, $10^{16}/\text{cm}^3$, while those on the *n*-side (p_n) may be $10^5/\text{cm}^3$ (if $N_A = 10^{16}/\text{cm}^3$ and $N_D = 10^{15}/\text{cm}^3$ in Si at room temperature). The holes would therefore be expected to diffuse so as to make their numbers more homogeneous throughout the material. Similar arguments hold for the electrons. Specifically, Fig. 2.2(d) and (e) indicate that the holes will diffuse from the *p*-side to the *n*-side and electrons will diffuse from the *n*-side to the *p*-side. Ideally the diffusion process would continue until

*The subscripts have been included to identify the majority and minority carriers; p_p means holes in a *p*-material, p_n means holes in *n*-material, etc.

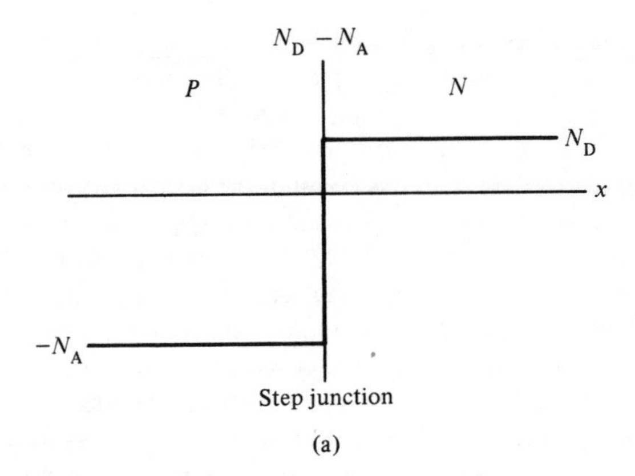

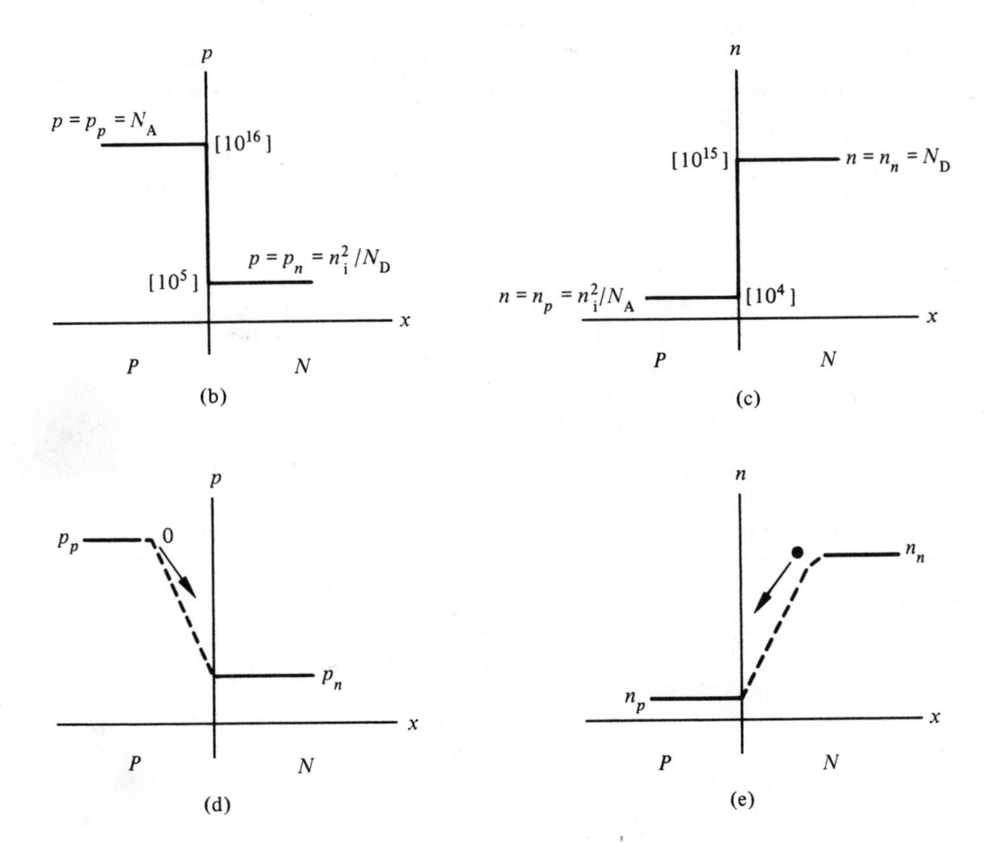

Fig. 2.2 (a) Step junction (b) and (c) hypothetical "initial" carrier concentrations; (d) and (e) approximate equilibrium carrier concentrations. Numbers in [ ] indicate typical values.

the carrier concentrations become equal on both sides of the junction. The diffusion process, however, cannot continue forever because it disrupts the charge balance between $-qN_A$ and qp_p on the p-side of the junction and between qN_D and $-qn_n$ on the n-side of the junction. When the holes diffuse away from the p-side, they leave behind the ionized acceptor atoms (N_A^-) that are fixed in place within the crystal lattice. On the n-side the electrons diffuse away, leaving the ionized donor atoms (N_D^+) and a charge density of qN_D. Obviously, a net charge density similar to Fig. 2.3(b) must be created by the reduction of the majority carrier concentrations. From Gauss's law, a net charge density implies in turn the existence of an electric field and therefore a potential difference. Since the charge is positive on the right-hand side of the junction and negative on the left-hand side of the junction, the electric field will be directed along the negative x-axis; that is, the electric field is negative. The electric field thus opposes diffusion of holes from the p-side and electrons from the n-side. Stated another way, "it acts to paste the carriers back in place" inhibiting further diffusion of the majority carriers.

The general form of the electric field can be established through the use of Eq. (2.1), where the net charge density (ρ) (coulombs/cm³) is equal to the imbalance between the charge carriers and the ions.

$$\mathscr{E}(x) = \frac{1}{K_S \varepsilon_0} \int_{-\infty}^{x} \rho(x)\, dx, \qquad (\text{V/cm}) \tag{2.1}$$

where

K_S = relative dielectric constant of the semiconductor (for Si $K_S = 11.8$)

$\varepsilon_0 = 8.854 \times 10^{-14}$, (farad/cm)

$$\rho(x) = q(p - n + N_D - N_A), \qquad (\text{coulombs/cm}^3) \tag{2.2}$$

Performing a rough graphical integration of Fig. 2.3(b), one obtains the electric field as sketched in Fig. 2.3(c).

The V. I. P. (very important point) of this section is that the carrier concentration difference between the n and p regions causes the carriers to diffuse. The diffusion, however, leads to a charge imbalance. The charge imbalance in turn produces an electric field, which counteracts the diffusion so that in thermal equilibrium the *net* flow of carriers is zero. The charged region near the metallurgical junction where the mobile carriers have been reduced is called the *depletion region*.

To continue our discussion of the qualitative aspects of the p-n junction, let us consider the potential, $V(x)$, within the depletion region and across the entire device. With a charge density and resultant electric field present within the structure, there must also be a potential gradient. From electromagnetic field theory,

$$\mathscr{E} = -\nabla V(x) \tag{2.3a}$$

or

$$\mathscr{E} = -dV/dx, \qquad \text{for one-dimensional problems} \tag{2.3b}$$

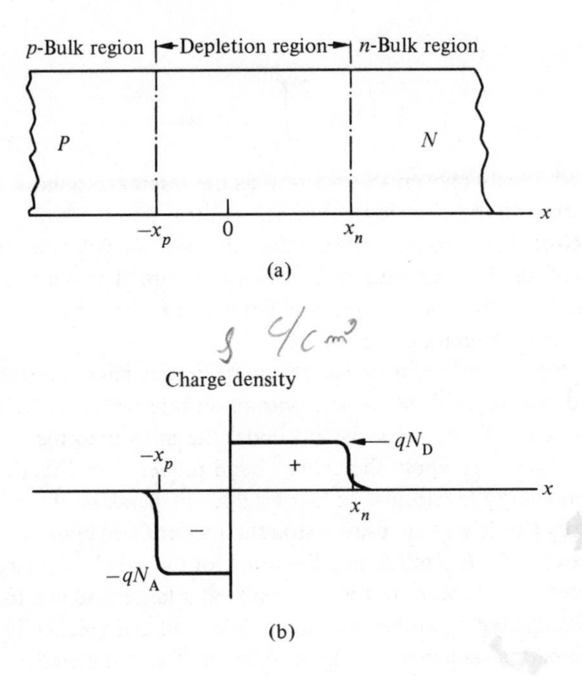

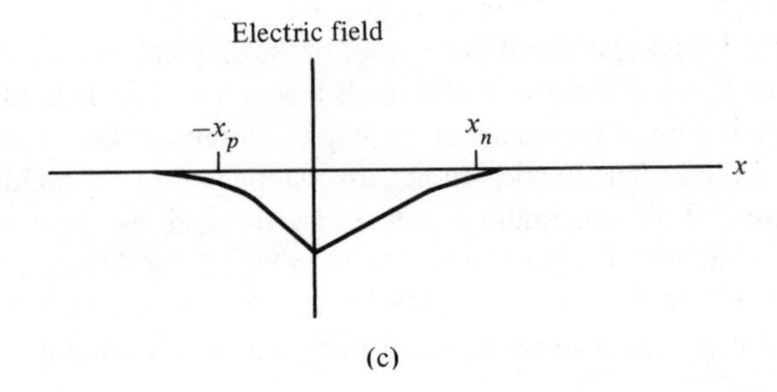

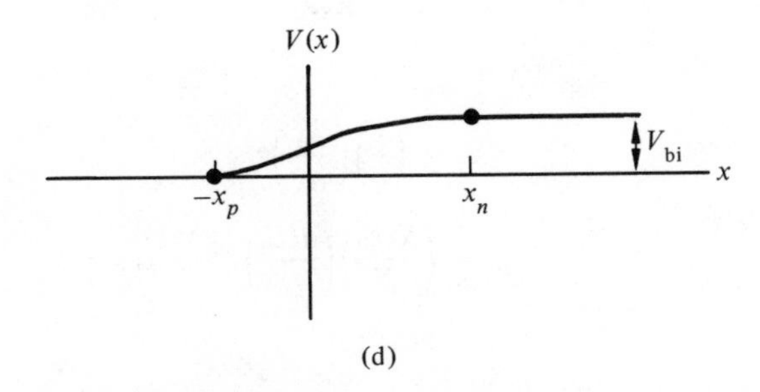

Fig. 2.3 Depletion region electrostatics.

and, integrating Eq. (2.3b),

$$V(x) = -\int_{-\infty}^{x} \mathscr{E}(x)\,dx \tag{2.4}$$

where we have arbitrarily chosen $V(-\infty) = 0$ as the reference point.* By performing a graphical integration on the electric field of Fig. 2.3(c), we obtain the potential function of Fig. 2.3(d). Note that a voltage (V_{bi}), called the *built-in potential*, exists across the depletion region of the device even at thermal equilibrium. The built-in potential can be thought of as similar to the contact potential between two dissimilar metals. In the next section we will derive a relationship for V_{bi}.

At this point the reader may well ask, "Can the energy band model be applied to the p-n junction and will it yield the same information relative to ρ, $\mathscr{E}$, $V(x)$, and V_{bi} as previously discussed?" To be a consistent model the answer to the second part of the question must be "yes." To apply the energy band model, recall that for thermal equilibrium the Fermi energy level must be a constant, independent of position. Therefore, to sketch the energy band diagram, draw a straight line for E_F on both sides of the junction. Next draw in lines for E_c, E_v, and E_i parallel to E_F for the p-side at a large distance from the junction. Repeat this procedure for the n-side at a large distance from the junction. The diagram is completed by connecting the conduction and valence band edges on the two sides of the junction such that E_G is kept constant. The completed energy band model is that of Fig. 2.4.

Several facts can be deduced from the energy band diagram. Remember from Volume I that the electric field is proportional to the slope of the diagram. Hence, from the negative slope of E_c, E_v, or E_i one concludes that there is a negative electric field in the depletion region. The slope is zero at the edges of the depletion region, indicating a zero electric field in the bulk regions. The maximum negative slope is near the middle, in agreement with the positioning of the maximum negative electric field, as illustrated previously in Fig. 2.3(c). Likewise, the energy difference qV_{bi} in Fig. 2.4 indicates a built-in potential difference (V_{bi}) between the ends of the device (the bulk regions). Finally, please note that the charge density can also be deduced from the energy band diagram. Since

$$\frac{d\mathscr{E}}{dx} = \frac{\rho}{K_S\varepsilon_0} \tag{2.5}$$

and

$$\mathscr{E} = \left(\frac{1}{q}\right)\left(\frac{dE_i}{dx}\right) \tag{2.6}$$

$$\rho = \left(\frac{K_S\varepsilon_0}{q}\right)\left(\frac{d^2E_i}{dx^2}\right) \tag{2.7}$$

---

*The potential is arbitrary to within a constant and therefore we could also have selected, for example, $V(\infty) = 0$ or $V(0) = 0$.

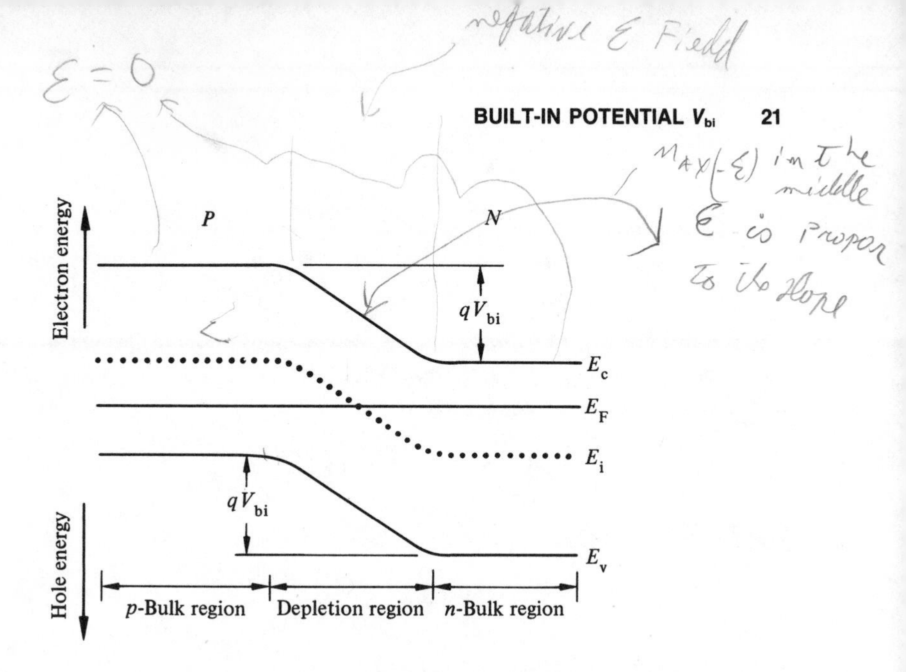

Fig. 2.4 A p-n junction energy band diagram at thermal equilibrium.

Thus the curvature (d^2E_i/dx^2) of the energy band diagram is proportional to the charge density. Clearly, the left-hand side of the depletion region with its negative curvature implies a negative charge density, while a positive curvature signifies the existence of a positive charge density in the n-material. One must conclude that the energy band model can be applied to the p-n junction and that it provides consistent information relative to the electric field, potential, and charge density within the diode.

2.2 BUILT-IN POTENTIAL V_{bi}

The previous section qualitatively established the existence of a built-in potential (V_{bi}) across the ends of the p-n junction diode from two viewpoints. The first viewpoint was the diffusion of carriers yielding a charge density, which in turn produced an electric field and hence the potential difference. The second and equivalent viewpoint was based on the energy band diagram and the constancy of E_F. Simply stated, the energy $E_C - E_F$ is different in the p and n regions and yields an energy difference qV_{bi}. In this section we establish an expression devoted to deriving an equation for V_{bi} that quantitatively relates V_{bi} to the doping difference between the p-region (N_A) and the n-region (N_D).

Thermal equilibrium $(V_A = 0)$ means that no net current flows. Stated more explicitly, $J_N = 0$, $J_P = 0$, and $J = 0$. With the net electron current zero, the current equation can be used to obtain the electric field. Setting the electron current equal to zero yields

$$J_N = J_{N|\text{drift}} + J_{N|\text{diffusion}} = q\mu n\mathscr{E} + qD_N\frac{dn}{dx} = 0 \tag{2.8}$$

Note that the drift current must be equal and opposite to the diffusion current, in order for Eq. (2.8) to be zero. Solving for the electric field yields

$$\mathscr{E} = \left(\frac{-qD_N}{q\mu_n n}\right)\left(\frac{dn}{dx}\right) = -\left(\frac{D_N}{\mu_n}\right)\left(\frac{1}{n}\right)\left(\frac{dn}{dx}\right) = -\left(\frac{kT}{q}\right)\left(\frac{1}{n}\right)\left(\frac{dn}{dx}\right) \tag{2.9}$$

The last form of Eq. (2.9) makes use of the Einstein relationship ("D ee over mu equals kT ee over q"). Using Eq. (2.4), the definition of potential, the voltage across the ends of the p-n junction can be written as

$$V_{\text{bi}} = -\int_{-\infty}^{\infty} \mathscr{E}\,dx = \frac{kT}{q}\int_{-\infty}^{\infty}\left(\frac{1}{n}\right)\left(\frac{dn}{dx}\right)dx = \frac{kT}{q}\int_{n(-\infty)}^{n(+\infty)}\frac{dn}{n} \tag{2.10}$$

Integrating, we obtain

$$V_{\text{bi}} = \frac{kT}{q}\ln n\Bigg|_{n(-\infty)}^{n(+\infty)} \tag{2.11}$$

Since, far from the junction on the p-side,

$$n_p = n(-\infty) = n_{\text{i}}^2/N_{\text{A}} \tag{2.12a}$$

and far from the junction on the n-side,

$$n_n = n(+\infty) = N_{\text{D}} \tag{2.12b}$$

we can write

$$V_{\text{bi}} = \frac{kT}{q}[\ln n_n - \ln n_p] = \frac{kT}{q}\ln\left[\frac{n_n}{n_p}\right] \tag{2.13}$$

or, by substituting Eqs. (2.12a) and (2.12b) into Eq. (2.13),

$$\boxed{V_{\text{bi}} = \frac{kT}{q}\ln\left[\frac{N_{\text{D}}N_{\text{A}}}{n_{\text{i}}^2}\right]} \tag{2.14}$$

Example: For silicon at room temperature ($kT = .026$ eV) doped $N_{\text{A}} = 10^{15}/\text{cm}^3$ on the p-side and $N_{\text{D}} = 10^{15}/\text{cm}^3$ on the n-side, where $n_{\text{i}} \simeq 10^{10}/\text{cm}^3$, then

$$V_{\text{bi}} = .026\ln\left[\frac{10^{15}\,10^{15}}{10^{20}}\right] = 0.599 \text{ volts}$$

If doped at $N_{\text{A}} = 10^{17}/\text{cm}^3$ and $N_{\text{D}} = 10^{15}/\text{cm}^3$, then V_{bi} calculates to 0.718 volts. Note that the greater the doping of either side, the greater V_{bi}. The reader can establish this fact from Eq. (2.14) or the energy band diagram of Fig. 2.4. If the p-side doping is increased, then E_{V} must move closer to E_{F} and qV_{bi} must increase. Doping the n-side to a higher degree moves E_{C} closer to E_{F}, also increasing qV_{bi}.

Silicon has a band gap of 1.12 eV, or, in terms of kT units at room temperature, $E_{\text{G}} = 43.08\ kT$. To keep the semiconductor from being degenerate, E_{F} must be equal to or greater than 3 kT from the conduction band edge or valence band edge. If the step junction were doped so that E_F were at 3 kT from the band edges on each side of the device, then, with the aid of Fig. 2.4,

$$qV_{bi} = 43.08\ kT - 6\ kT = 0.9641\ \text{eV}$$

and therefore

$$V_{bi} = 0.9641\ \text{volts}$$

is the maximum value of V_{bi} at room temperature without having degenerate silicon.

2.3 THE DEPLETION APPROXIMATION

The quantitative solutions for the charge density, $\mathscr{E}$, and $V(x)$ across the p-n junction under thermal equilibrium conditions are centered around the solution of Poisson's equation, repeated here for the reader's convenience.

$$\frac{d\mathscr{E}}{dx} = \frac{q}{K_S\varepsilon_0}(p - n + N_D - N_A) \tag{2.15a}$$

or

$$\frac{d^2V}{dx^2} = -\frac{q}{K_S\varepsilon_0}(p - n + N_D - N_A) \tag{2.15b}$$

In general $\mathscr{E}$, V, p, n, N_D, and N_A are functions of x, except for uniform doping where N_D and N_A are constants. Poisson's equation in its exact form is not easily solved for most devices because p and n are in turn functions of V and $\mathscr{E}$, the unknowns. To solve* the equation in "closed form" it is necessary to make several simplifications. A "closed-form" solution results in an equation for V as an explicit function of x. A particularly useful set of simplifications that allow the solution of Poisson's equation to be obtained explicitly are collectively called the *depletion approximation*. This approximation is justified if the solution can explain experimental data obtained from a real diode, and it can.

Figure 2.5(b) illustrates the charge density for the step junction as developed in the previous section. The depletion approximation assumes that the mobile carriers (n and p) are small in number compared to the donor and acceptor ion concentrations in the depletion region, and that the device is charge-neutral elsewhere. Stated mathematically and specialized to the uniformly doped p-n step junction:

2.3.1 Depletion Approximation

1. $N_A \gg n_p$ or p_p, hence $\rho = -qN_A^-$ for $-x_p \le x \le 0$.
2. $N_D \gg n_n$ or p_n, hence $\rho = qN_D^+$ for $0 \le x \le x_n$.
3. The charge density is zero in the bulk regions; that is, for $x > x_n$ and $x < -x_p$.

---

*One can numerically solve the equation on a large digital computer by making successive guesses until an iterative solution is obtained at each point throughout the device.

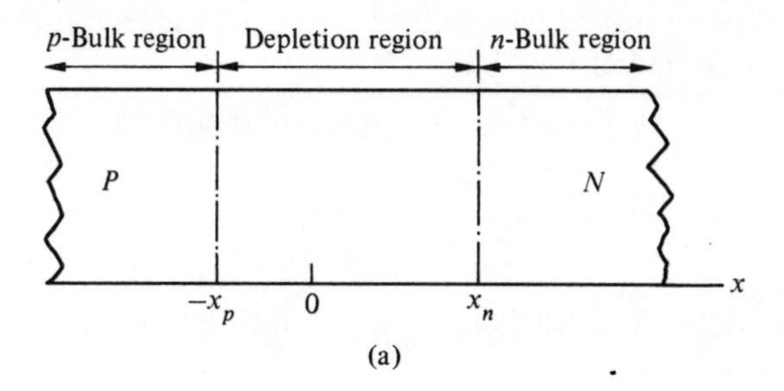

(a)

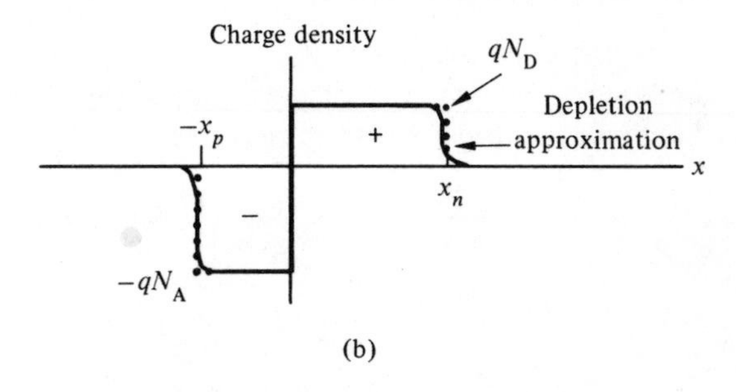

(b)

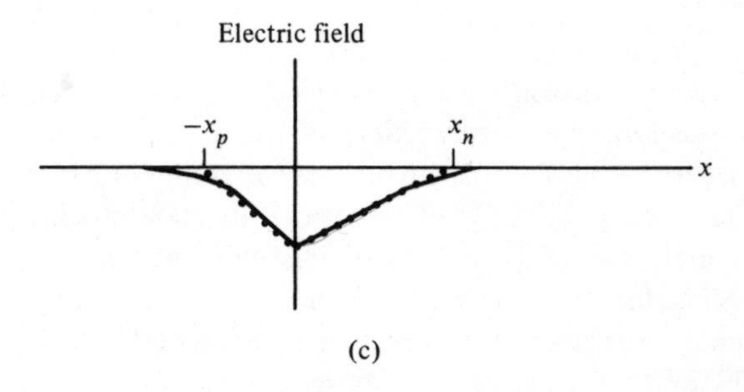

(c)

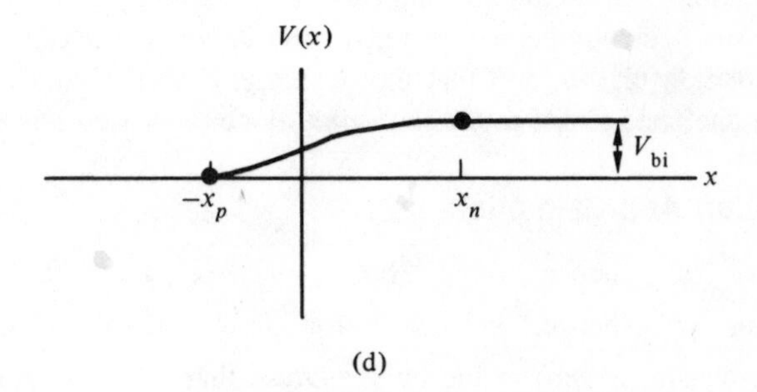

(d)

Fig. 2.5 Depletion approximation to the step junction.

The depletion region is bounded by $-x_p$ and x_n, while the regions outside the depletion region are called the n- and p-bulk regions, respectively.

The depletion approximation, with the appropriate charge density distribution as presented in Fig. 2.5, reduces Poisson's equation to

$$\frac{d\mathscr{E}}{dx} = \frac{qN_D}{K_S\varepsilon_0}, \qquad \text{for } 0 \le x \le x_n \tag{2.16a}$$

and

$$\frac{d\mathscr{E}}{dx} = \frac{-qN_A}{K_S\varepsilon_0}, \qquad \text{for } -x_p \le x \le 0 \tag{2.16b}$$

where

$\mathscr{E} = 0$ in the bulk regions and at the edges of the depletion region.

Solution for $\mathscr{E}$. The electric field can be obtained for the one-dimensional step junction by simply integrating Eqs. (2.16a) or (2.16b), from where one knows the electric field (at x_n and $-x_p$) to an arbitrary point inside the depletion region. Consider the p-side depletion region and Eq. (2.16b), remembering that $\mathscr{E}(-x_p) = 0$,

$$d\mathscr{E} = \frac{-qN_A}{K_S\varepsilon_0}\,dx \tag{2.17}$$

$$\mathscr{E}(x) = \int_0^{\mathscr{E}(x)} d\mathscr{E} = \int_{-x_p}^{x} \left[\frac{-qN_A}{K_S\varepsilon_0}\right] dx = \frac{-qN_A}{K_S\varepsilon_0}\, x \Big|_{x_p}^{x} \tag{2.18}$$

$$\mathscr{E}(x) = \frac{-qN_A}{K_S\varepsilon_0}(x_p + x), \qquad \text{for } -x_p \le x \le 0 \tag{2.19}$$

Note that the electric field is negative in agreement with our qualitative discussions. Also note that $\mathscr{E}(x)$ is the equation of a straight line with a negative slope.

The n-depletion region is obtained by integrating Eq. (2.16a) with $\mathscr{E}(x_n) = 0$. Starting at some value of x greater than zero and integrating to x_n

$$\mathscr{E}(x) = \int_{\mathscr{E}(x)}^{0} d\mathscr{E} = \int_{x}^{x_n} \frac{qN_D}{K_S\varepsilon_0}\,dx \tag{2.20}$$

$$-\mathscr{E}(x) = \frac{qN_D}{K_S\varepsilon_0}\, x \Big|_{x}^{x_n} = \frac{qN_D}{K_S\varepsilon_0}(x_n - x) \tag{2.21}$$

or

$$\mathscr{E}(x) = \frac{-qN_D}{K_S\varepsilon_0}(x_n - x), \qquad \text{for } 0 \le x \le x_n \tag{2.22}$$

Note that Eq. (2.22) is a straight line with positive slope. Eqs. 2.19 and 2.22 are plotted in Fig. 2.5(c) as the dotted line.

$\mathcal{E}_P(0) = \mathcal{E}_N(0)$

As indicated in Fig. 2.5(c), the electric field must be continuous at $x = 0$ since there is no layer of charge at that position. Thus from Eqs. (2.19) and (2.22), evaluated at $x = 0$,

$$\frac{-qN_A}{K_S\varepsilon_0}x_p = \frac{-qN_D}{K_S\varepsilon_0}x_n \tag{2.23}$$

or

$$[\![N_A x_p = N_D x_n]\!] \tag{2.24}$$

Equation (2.24) multiplied by qA (where A is diode area) states that the total negative charge must equal the total positive charge. The equality is also evidenced in Fig. 2.5(b) where the area of the plot x_pN_A must be equal to the area of x_nN_D. In Fig. 2.5 $N_A > N_D$; hence $x_p < x_n$—an easy method of relating the doping to depletion widths.

Solution for V(x). The potential function $V(x)$ within the depletion region is derived from the definition of potential; that is,

$$\frac{dV}{dx} = -\mathscr{E} \tag{2.25}$$

Combining Eq. (2.25) with the Eq. (2.19) expression for the electric field in the p-side depletion region yields

$$\frac{dV}{dx} = \frac{qN_A}{K_S\varepsilon_0}(x_p + x) \tag{2.26}$$

Separating variables and integrating, gives

$$V(x) = \int_0^{V(x)} dV = \frac{qN_A}{K_S\varepsilon_0}\int_{-x_p}^{x}(x_p + x)\,dx \tag{2.27}$$

where we have arbitrarily selected the reference potential to be zero in the bulk p-region; that is $V(-x_p) = 0$. Completing the integration yields

$$V(x) = \frac{qN_A}{2K_S\varepsilon_0}(x_p + x)^2; \qquad \text{for } -x_p \le x \le 0 \tag{2.28}$$

Note that $V(x)$ is a parabola of positive curvature and is displaced down the negative x-axis by x_p units.

For the n-depletion region

$$\frac{dV}{dx} = -\mathscr{E} = \frac{qN_D}{K_S\varepsilon_0}(x_n - x) \tag{2.29}$$

Since the bulk p-region was chosen to be at zero potential, the bulk n-region must be at V_{bi}; that is, $V(x_n) = V_{bi}$. Integrating,

$$\int_{V(x)}^{V_{bi}} dV = V_{bi} - V(x) = \frac{qN_D}{K_S\varepsilon_0}\int_x^{x_n}(x_n - x)\,dx = \frac{-qN_D}{2K_S\varepsilon_0}\left(x_n x - \frac{x^2}{2}\right)\Bigg|_x^{x_n} \tag{2.30}$$

$$V(x) = \frac{-qN_D}{2K_S\varepsilon_0}(x_n - x)^2 + V_{bi}, \qquad \text{for } 0 \le x \le x_n \tag{2.31}$$

Figure 2.5(d) illustrates the potential function $V(x)$ for the p-n step junction. By comparing Figs. 2.5(d) and 2.4, one can see that the potential function is the horizontal mirror image of the energy band diagram E_c, E_v, or E_i.

2.3.2 Depletion Region Width

The $\mathscr{E}(x)$ and $V(x)$ functions for the p-n junction are written in terms of the parameters N_A, N_D, x_n, and x_p. Here N_D and N_A are known from resistivity measurements. The reader should be asking the question, "How are the depletion distances x_n and x_p related to the material parameters, especially those parameters that could be measured?" To derive such a relationship, the most promising starting point is Eq. (2.31), since all the parameters except x_n are known. Remember V_{bi} was written in terms of N_A and N_D in Eq. (2.14). Since there is no dipole layer at $x = 0$, the potential function must be continuous; that is, $V(0^-) = V(0^+)$. With the aid of Eqs. (2.28) and (2.31), evaluated at $x = 0$, we conclude

$$\left[\frac{qN_A}{2K_S\varepsilon_0}\right]x_p^2 = \left[\frac{-qN_D}{2K_S\varepsilon_0}\right]x_n^2 + V_{bi} \tag{2.32}$$

Equations (2.32) and (2.24) constitute two equations and two unknowns. Solving for x_p from Eq. (2.24)

$$x_p = \left[\frac{N_D}{N_A}\right]x_n \tag{2.33}$$

and substituting into Eq. (2.32),

$$\frac{qN_A}{2K_S\varepsilon_0}\frac{N_D^2}{N_A^2}x_n^2 = \frac{qN_D^2x_n^2}{2K_S\varepsilon_0N_A} = \frac{-qN_D}{2K_S\varepsilon_0}x_n^2 + V_{bi} \tag{2.34}$$

We can now solve for x_n^2,

$$x_n^2 = \frac{2K_S\varepsilon_0}{q}[V_{bi}]\frac{1}{[(N_D^2/N_A) + N_D]} \tag{2.35}$$

or

$$x_n = \left[\frac{2K_S\varepsilon_0V_{bi}}{q}\frac{N_A}{N_D(N_A + N_D)}\right]^{1/2} \tag{2.36}$$

Similarly for x_p,

$$x_p = \left[\frac{2K_S\varepsilon_0 V_{\text{bi}}}{q} \frac{N_D}{N_A(N_A + N_D)}\right]^{1/2} \tag{2.37}$$

The *depletion width* (W) can be determined from x_n and x_p to be

$$W = x_p + x_n \tag{2.38}$$

$$W = \left[\frac{2K_S\varepsilon_0 V_{\text{bi}}}{q(N_A + N_D)}\right]^{1/2} \left[\sqrt{\frac{N_A}{N_D}} + \sqrt{\frac{N_D}{N_A}}\right] \tag{2.39}$$

With some algebraic manipulation, Eq. (2.39) can be written in a slightly more convenient form:

$$W = \left[\frac{2K_S\varepsilon_0 V_{\text{bi}}}{q} \frac{(N_A + N_D)}{N_A N_D}\right]^{1/2} \tag{2.40}$$

Example:

$$kT = .026\,\text{eV},$$

$$N_A = 10^{16}/\text{cm}^3,\ N_D = 10^{15}/\text{cm}^3, \qquad V_{\text{bi}} = \frac{kT}{q} \ln\left[\frac{10^{16}10^{15}}{(10^{10})^2}\right] = 0.659 \text{ volts},$$

$$W = \left[\frac{2 \times 11.8 \times 8.854 \times 10^{-14} \times 0.659}{1.6 \times 10^{-19}} \times \frac{(10^{16} + 10^{15})}{10^{16} \times 10^{15}}\right]^{1/2},$$

$$W = 0.9730 \times 10^{-4} \text{ cm} \quad \text{or} \quad 0.973 \text{ microns}$$

$$x_n = \left[\frac{2 \times 11.8 \times 8.854 \times 10^{-14} \times 0.659}{1.6 \times 10^{-19}} \times \frac{10^{16}}{10^{15}(10^{16} + 10^{15})}\right]^{1/2}$$

$$x_n = 0.88455 \times 10^{-4} \text{ cm} \quad \text{or} \quad 0.88455 \text{ microns}$$

Similarly from Equation (2.37),

$$x_p = 0.08845 \times 10^{-4} \text{ cm} \quad \text{or} \quad 0.08845 \text{ microns}$$

Note that since the p-side is doped greater, $x_n > x_p$; that is, the junction *depletes further into the lighter-doped material*.

We remind the reader that the discussion and equations derived thus far in the chapter are for thermal equilibrium. The next section presents the case of forward bias, $V_A > 0$, and then reverse bias, $V_A < 0$.

2.4 ELECTROSTATICS OF FORWARD AND REVERSE BIAS

The electric field, charge density, and potential function for the case of positive applied voltage ($V_A > 0$) could be developed with discussions and derivations parallel with the previous two sections. However, there is an easier approach. Consider Fig. 2.6(a), where the diode is at thermal equilibrium. No current flows and there is no voltage drop or electric field in the bulk p- and n-regions. The *ohmic* contacts of the metal-semiconductor junctions have *contact potentials* V_P and V_N that are fixed in value and depend only on the materials used to make the device. The *junction voltage* (V_j) appears across the edges of the depletion region. At thermal equilibrium $V_j = V_{bi}$. Since $V_A = 0$, writing a loop equation yields

$$V_j = V_N - 0 + V_P = V_{bi} \tag{2.41}$$

and

$$V_{bi} = V_N + V_P \tag{2.42}$$

is a fixed value depending only on the doping of the semiconductor.

The diode is forward biased when the applied voltage (V_A) has a positive potential on the p-region and a negative potential on the n-region as illustrated in Fig. 2.6(b). Since V_A is opposite in polarity relative to V_j, it must reduce the voltage across the depletion region. Writing a loop equation, we have

$$V_j = V_N - V_A + V_P = V_N + V_P - V_A \tag{2.43}$$

and substituting from Eq. (2.42),

$$[\![V_j = V_{bi} - V_A]\!] \tag{2.44}$$

under the assumption that no voltage drop occurs in the bulk p- and n-regions.

In the previous section, for thermal equilibrium the junction potential was V_{bi} and this was used as a boundary condition in the solution of Poisson's equation. All that needs to be changed to obtain a solution for the case of an applied voltage is the replacement of V_{bi} by $(V_{bi} - V_A)$.

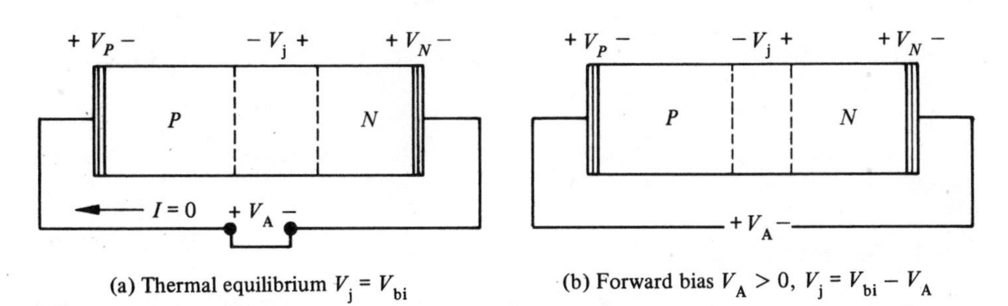

(a) Thermal equilibrium $V_j = V_{bi}$ (b) Forward bias $V_A > 0$, $V_j = V_{bi} - V_A$

Fig. 2.6 Junction potential: (a) thermal equilibrium, $V_j = V_{bi}$; (b) forward bias $V_A > 0$, $V_j = V_{bi} - V_A$.

2.4.1 The n-depletion Region, $0 \leq x \leq x_n$

$$x_n = \left[\frac{2K_S\varepsilon_0}{q}(V_{bi} - V_A)\frac{N_A}{N_D(N_A + N_D)}\right]^{1/2} \tag{2.45}$$

$$V(x) = (V_{bi} - V_A) - \frac{qN_D}{2K_S\varepsilon_0}(x_n - x)^2 \tag{2.46}$$

$$\mathscr{E}(x) = \frac{-qN_D}{K_S\varepsilon_0}(x_n - x) \tag{2.47}$$

2.4.2 The p-depletion Region, $-x_p \leq x \leq 0$

$$x_p = \left[\frac{2K_S\varepsilon_0}{q}(V_{bi} - V_A)\frac{N_D}{N_A(N_A + N_D)}\right]^{1/2} \tag{2.48}$$

$$V(x) = \frac{qN_A}{2K_S\varepsilon_0}(x_p + x)^2 \tag{2.49}$$

$$\mathscr{E}(x) = \frac{-qN_A}{K_S\varepsilon_0}(x_p + x) \tag{2.50}$$

and

$$W = \left[\frac{2K_S\varepsilon_0}{q}(V_{bi} - V_A)\left(\frac{N_A + N_D}{N_A N_D}\right)\right]^{1/2} \tag{2.51}$$

2.4.3 Forward Bias, $V_A > 0$

Consider the results of forward bias as compared to thermal equilibrium. Since $(V_{bi} - V_A)$ is less than V_{bi} for thermal equilibrium, x_n and x_p are reduced, as illustrated in Fig. 2.7. Also shown are the effects of $V_A > 0$ on the potential, electric field, and width of the charge density; each is reduced. This may be deduced from Eqs. (2.45) through (2.51), which reduce to the thermal equilibrium relationships when $V_A = 0$. An important point about forward bias is that, for these equations to be valid,

$$V_A < V_{bi}.$$

If not, Kirchhoff's Voltage Law is violated. The only way to remove this restriction is to allow for voltage drops in the bulk regions.

2.4.4 Reverse Bias, $V_A < 0$

The case of reverse bias requires that the applied voltage be less than zero, that is, of opposite polarity to forward bias. By inspection of Eq. (2.44), we note that the applied voltage "adds" to V_{bi} and thereby increases V_j. Since V_A is always a negative number, it is therefore always less than V_{bi}—a nice result which allows the use of Eqs. (2.45) through (2.51) directly without any concern for the magnitude of V_A.

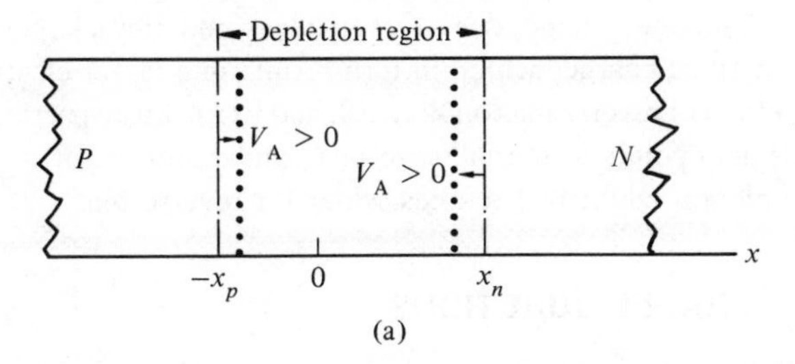

(a)

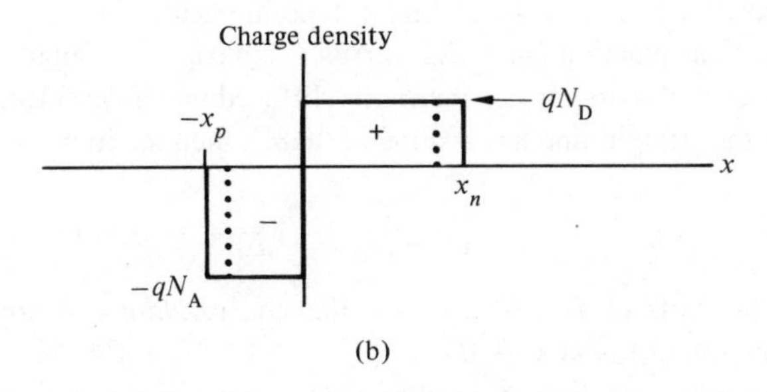

(b)

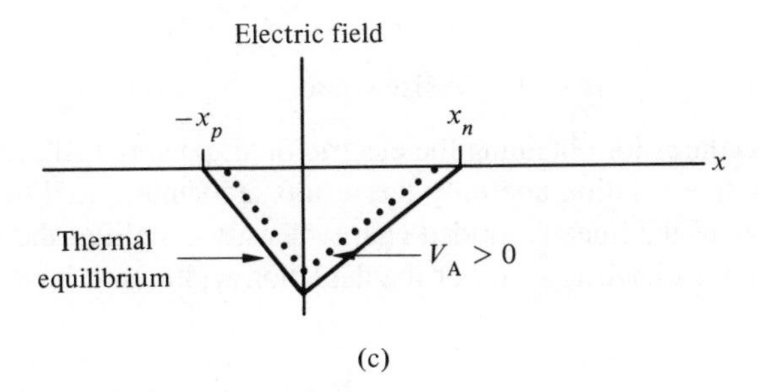

(c)

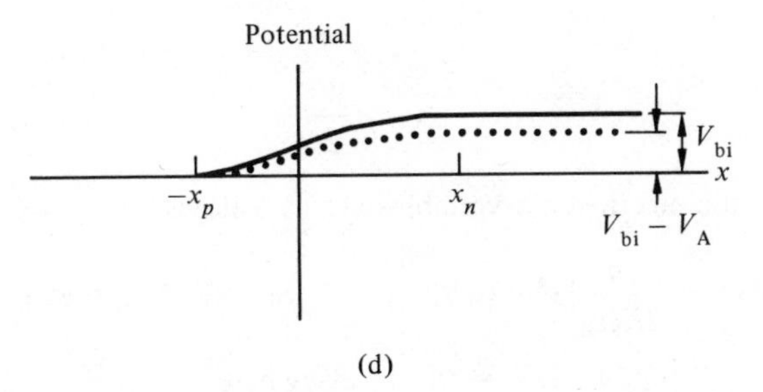

(d)

Fig. 2.7 Effect of forward bias on the diode electrostatics [$V_A > 0$, dotted lines; $V_A = 0$, unbroken lines].

The increase in junction voltage, due to reverse bias, requires a larger depletion width to produce a larger fixed charge, which in turn results in a larger electric field. Figure 2.8 illustrates the effect of reverse bias on W, ρ, $\mathscr{E}$, and $V(x)$. An inspection of Eqs. (2.45), (2.48), and (2.51) also points to the increase in x_n and x_p for larger negative values of V_A; that is, the depletion width W becomes wider for reverse bias.

2.5 LINEARLY GRADED JUNCTIONS

The *ideal linearly graded* junction is an approximation to the situation where impurities are thermally diffused or ion-implanted into a semiconductor to form a p-n junction. Figure 1.5(a) illustrated the case of adding n-type impurities into a p-substrate, where near the metallurgical junction $N_D - N_A$ versus x is nearly a straight line (linear).

Consider the case of p-impurities thermally diffused into an n-substrate as illustrated in Fig. 2.9(a). If the straight line has a slope of "$-a$", then the impurity distribution can be described by

$$N_A(x) - N_D = -ax \tag{2.52}$$

where "a" has the units of $\#/\text{cm}^4$ and is called the *grading constant*. Note that the metallurgical junction (x_j) is at $x = 0$.

The depletion approximation, as applied to the graded junction, is illustrated in Fig. 2.9(b). In particular, for the charge density,

$$\begin{aligned} \rho &= qax, \qquad \text{for } -x_p \le x \le x_n \\ \rho &= 0, \qquad \text{elsewhere} \end{aligned} \tag{2.53}$$

The solution procedures for obtaining the electric field, potential, W, and V_{bi} are similar to the abrupt-junction solution and only the results are summarized here.

The symmetry of the linearly graded charge density simplifies the solution because by inspection we must have $x_n = x_p$, or the depletion width must be symmetrical about x_j. Thus

$$x_n = \frac{W}{2} \tag{2.54a}$$

and

$$x_p = \frac{W}{2} \tag{2.54b}$$

Relationships for the electrostatic variables are as follows:

$$\begin{aligned} \mathscr{E}(x) &= \frac{qa}{2K_S\varepsilon_0}[x^2 - (w/2)^2], \qquad \text{for } -W/2 \le x \le W/2 \\ \mathscr{E} &= 0, \qquad \text{elsewhere} \end{aligned} \tag{2.55}$$

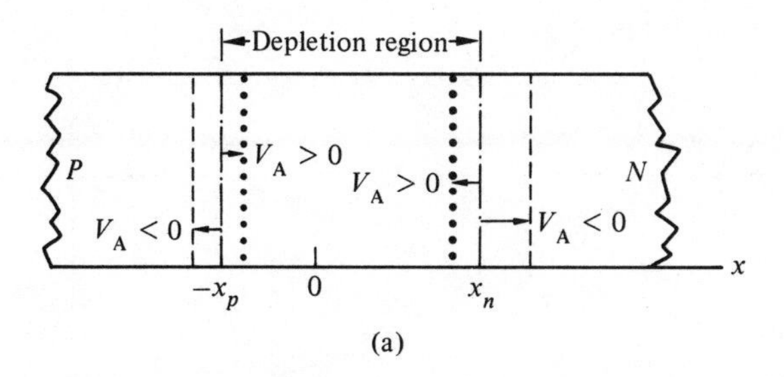

(a)

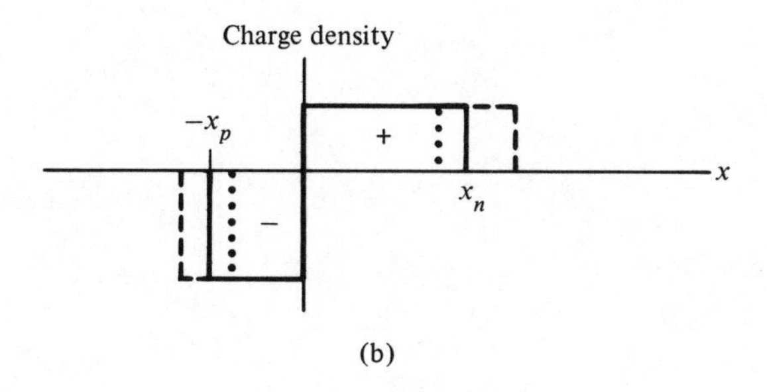

(b)

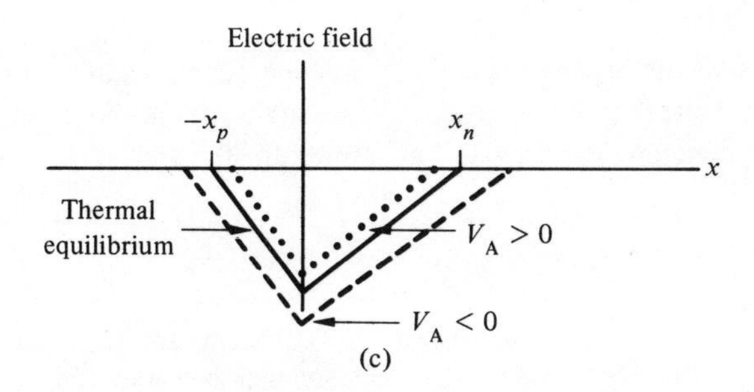

(c)

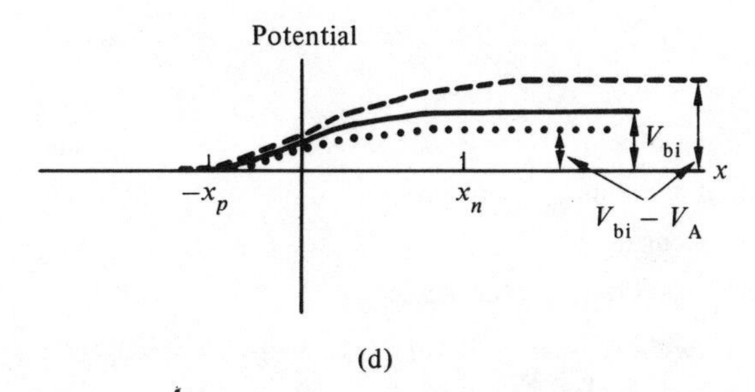

(d)

Fig. 2.8 Effect of bias on depletion region electrostatics.

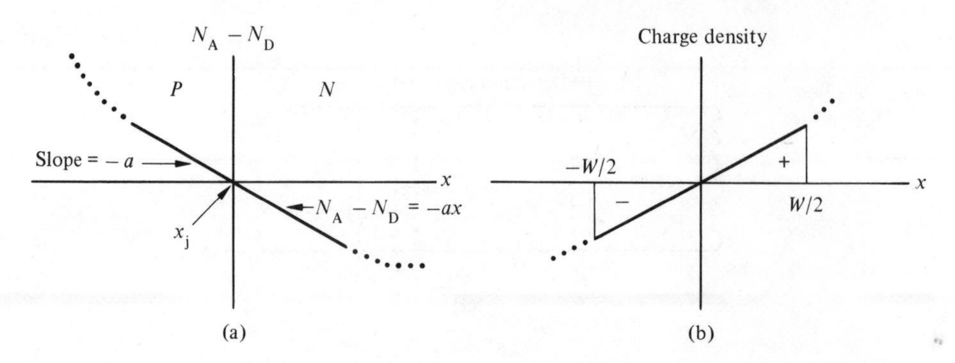

Fig. 2.9 Linear graded p-n junction.

$$V(x) = \frac{qa}{6K_S\varepsilon_0}\left[2\left(\frac{W}{2}\right)^3 + 3\left(\frac{W}{2}\right)^2 x - x^3\right], \qquad \frac{-W}{2} \le x \le \frac{W}{2} \tag{2.56}$$

$$W = \left[\frac{12K_S\varepsilon_0}{qa}(V_{bi} - V_A)\right]^{1/3} \tag{2.57}$$

$$V_{bi} = \frac{2kT}{q}\ln\left[\frac{aW}{2n_i}\right], \qquad W \text{ at } V_A = 0 \tag{2.58}$$

The case of forward and reverse bias applies to these equations in the same manner as for the step junction. Note that W increases or decreases as the one-third power for the linearly graded junction compared to the one-half power for the step junction.

PROBLEMS

2.1 A Si step junction has a doping of $N_A = 5 \times 10^{15}/\text{cm}^3$ and $N_D = 10^{15}/\text{cm}^3$ and a cross-sectional area (A) of 10^{-4} cm^2. Assume the depletion approximation and

(a) Calculate V_{bi}.

(b) Calculate x_n, x_p and W.

(c) What is the total positive ion charge?

(d) Calculate the electric field at $x = 0$.

(e) Sketch to scale ρ and $\mathscr{E}$ in the depletion region.

(f) n_{p0} and p_{n0} are calculated as?

(g) Draw the energy band diagram for the device.

2.2 A Si step junction with $N_A = 4 \times 10^{18}/\text{cm}^3$ (assumed to be nondegenerate) and $N_D = 10^{16}/\text{cm}^3$ is at room temperature. Calculate

(a) V_{bi}.

(b) x_n, x_p and W.

(c) Electric field at $x = 0$.

(d) Sketch ρ and $\mathscr{E}$ to scale.

(e) n_{p0} and p_{n0}.

(f) Draw the energy band diagram for the diode.

2.3 A Si p^+-n junction diode has a doping of $N_A = 10^{17}/\text{cm}^3$ on the p-side and $N_D = 10^{15}/\text{cm}^3$ on the n-side. Assuming room temperature and equilibrium conditions,

(a) Determine V_{bi} (the built-in potential).

(b) Determine x_n, x_p and W.

(c) *Plot* the electrostatic potential (V) as a function of position inside the diode. (Note critical point values of V and x on your plot.)

(d) *Plot* the electric field ($\mathscr{E}$) as a function of position inside the diode. (Note critical point values of $\mathscr{E}$ and x on your plot.)

(e) Draw the energy band diagram for the diode.

(f) As a general rule, the physical state and characteristics of a p-n junction, where the doping on one side is much greater than the doping on the other side, is controlled by the parameters of the lightly doped side. To what extent is this "rule" followed in this particular problem? Discuss.

2.4 A step junction Si diode maintained at room temperature is doped such that $E_F = E_V - 2kT$ on the p-side and $E_F = E_C - E_G/4$ on the n-side. $A = 10^{-3}\text{cm}^2$,

(a) Draw the equilibrium energy band diagram for the diode.

(b) What is the built-in voltage (V_{bi}) that exists across the diode under equilibrium conditions? Give both a symbolic and numerical answer.

(c) State briefly what is meant by the "depletion approximation" used in the analysis of semiconductor devices.

2.5 Silicon is doped as a p_1-p_2 step junction in Fig. P2.5, where $N_{A_1} < N_{A_2}$. (This type of junction is often called an isotype junction because of the same type of doping on both sides.)

(a) Sketch the energy band diagram (N_{A_1} is on the left).

(b) Derive an expression for V_{bi}.

(c) Based on the energy band diagram sketch what the $\rho(x)$ and $\mathscr{E}(x)$ plots might approximate.

(d) Same as (c) for $V(x)$.

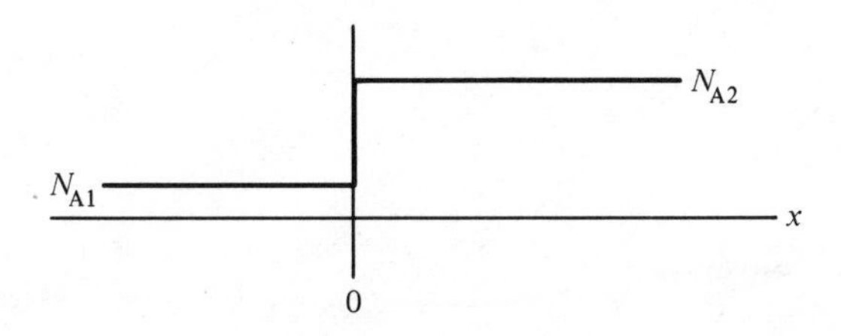

Figure P2.5

2.6 For the linearly graded junction assume V_{bi} is known and derive an equation for

(a) Electric field, Eq. (2.55).

(b) $V(x)$, Eq. (2.56).

(c) W, Eq. (2.57).

3 / The Diode Volt–Ampere Characteristic

The depletion region electrostatics presented in Chapter 2 are first extended to include a visualization of carrier flow (particle flux) taking place within the device under equilibrium conditions. With the basic current-flow models firmly established, the qualitative form of the volt–ampere (V–I) characteristic is then deduced from a consideration of the energy band diagrams and carrier fluxes under forward and reverse biases. A first-order quantitative relationship for the V–I dependence, the ideal diode equation, is next derived after establishing a "game plan" for solving the bulk p- and n-region "equations of state." In discussing the "game plan," special consideration is given the carrier densities at the edges of the depletion region. These densities are needed as boundary conditions in the analytical derivation. The analytical derivation itself yields the minority carrier concentrations, the carrier currents, and the total diode current in terms of the externally applied voltage (V_A). Much of the development in this chapter makes use of the depletion approximation and assumes a step junction device.

3.1 THERMAL EQUILIBRIUM

The diode at thermal equilibrium serves as a base upon which to build the concepts of carrier flux and potential barriers. Indeed, when forward or reverse biases are applied, the changes in the potential barrier and carrier fluxes determine the current direction and its relative magnitude.

The energy band diagram under equilibrium conditions is illustrated in Fig. 3.1. Since E_F is a constant at thermal equilibrium, the band edges (E_c and E_v) must change their position relative to E_F when making the transition from p- to n-material. As discussed previously in Chapter 2, the slope of the energy band edges is proportional to the electric field; in this case a negative slope yields a negative electric field in the depletion region. The carrier pyramids in the bulk regions of the figure crudely represent the energy distribution of the carrier densities as determined by the product of the density of states function and Fermi function. The majority and minority carriers are represented by the relative number of appropriate symbols. It should be kept in mind that the carrier

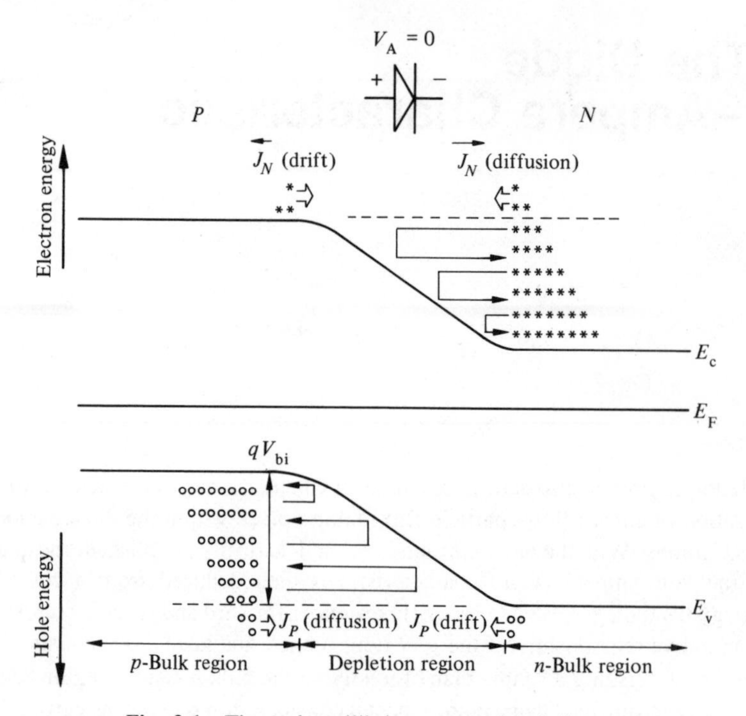

Fig. 3.1 Thermal equilibrium energy band diagram.

concentrations are actually decreasing exponentially with increasing energy and that the figure is only a model.

The reader might well ask, "Are the carriers in or near the depletion region edges drifting down the 'potential hill'?" That is, are minority electrons in the p-material drifting down the energy band diagram from left to right and holes in the n-type material drifting from right to left to seek a lower energy? This implies a current flow, which cannot be correct for thermal equilibrium. The key to answering this quandary is that the *net current (or carrier flux) must be zero for each carrier type*. Holes cannot build up at one end of the device!

Consider the two ways in which current can flow: drift in an electric field and diffusion due to a concentration gradient. The current equations are repeated here for the reader's convenience.

$$\mathbf{J}_P = \mathbf{J}_{P|\text{drift}} + \mathbf{J}_{P|\text{diffusion}} \tag{3.1}$$

$$J_P = q\mu_p p\mathscr{E} - qD_P\frac{dp}{dx}$$

$$\mathbf{J}_N = \mathbf{J}_{N|\text{drift}} + \mathbf{J}_{N|\text{diffusion}} \tag{3.2}$$

$$J_N = q\mu_n n\mathscr{E} + qD_N \frac{dn}{dx}$$

$$J = J_P + J_N \tag{3.3}$$

For thermal equilibrium, $J = J_P = J_N = 0$. A glance at Eqs. (3.1) and (3.2) quickly reveals the need for

$$J_{N|\text{drift}} = -J_{N|\text{diffusion}}$$

and

$$J_{P|\text{drift}} = -J_{P|\text{diffusion}}$$

Consequently, in answer to our quandary, the zero net current (or carrier flux) for each carrier type under equilibrium conditions is achieved through a cancellation of the drift component by an oppositely directed diffusion component of equal magnitude.

Expanding on the foregoing conclusion, examine the carrier concentrations on the two sides of the junction. The majority carrier p_p may be, for example, $10^{16}/\text{cm}^3$, as compared to p_n which may be $10^5/\text{cm}^3$. In going from the p- to the n-side the hole concentration has changed by eleven orders of magnitude! A similar case holds for electrons (see Fig. 3.2). Since, as discussed in Chapter 2, a typical depletion width may be $\sim 10^{-4}$cm, therefore dp/dx is very large, causing the diffusion of holes from left to right in Fig. 3.1. However, the diffusing holes must climb the "potential hill," qV_{bi}, in order to enter the n-bulk region. The electric field (and hence potential) tries to "paste them back." Figure 3.1 illustrates this as the holes being reflected back to the p-region. Only those holes with energies greater than qV_{bi} can diffuse into the n-region and constitute the diffusion component of the hole current, a positive current.

Consider the minority carrier holes generated in the n-region near the depletion region edge. These holes can "fall down" the potential hill since holes like to "float" in the energy band diagram; that is, they *drift* in the electric field from right to left. This constitutes a negative hole current component.

Summarizing, at thermal equilibrium, in the depletion region, the hole current components are not individually zero, that is

$$J_{P|\text{drift}} \neq 0 \text{ and } J_{P|\text{diffusion}} \neq 0$$

but $J_P = 0$ because the components are *equal in magnitude but oppositely directed*. Similar arguments hold true for electrons as indicated in Fig. 3.1. The carrier flux due to drift is equal and opposite to that of diffusion, yielding a *net* electron flux of zero. Thus, the net electron current is zero, with the drift and diffusion current components equal in magnitude, but oppositely directed.

Finally, in the uniformly doped bulk regions, $E_c - E_F$ is a constant and hence the electric field is zero. However, since $n = N_D$ = constant and $p = N_A$ = constant, dp/dx and dn/dx are zero. Therefore the drift and diffusion currents (or particle fluxes) for electrons and holes are each zero and the total current is zero.

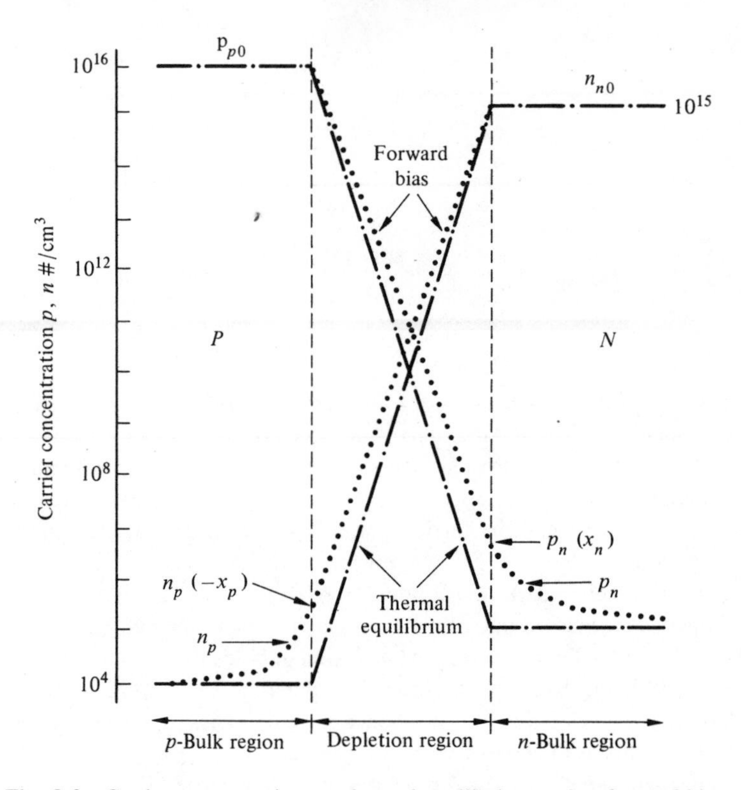

Fig. 3.2 Carrier concentrations at thermal equilibrium and at forward bias.

3.2 QUALITATIVE $V-I$ CHARACTERISTICS

This section extends the carrier flux energy band diagram discussion of the previous section to describe the qualitative nature of the volt–ampere characteristic for the diode. In particular it explains why the current flow is large in one direction, but very small in the opposite direction.

3.2.1 Forward Bias, $V_A > 0$

The effect of applying a negative potential to the n-region (with respect to the p-region) on the energy band diagram is illustrated in Fig. 3.3. The p-side was arbitrarily chosen to remain fixed in position, with the n-side bulk region being shifted upward in electron energy by qV_A. Note that the slope of the band edges in the depletion region is reduced in magnitude as compared to thermal equilibrium. This means that the magnitude of the electric field has been reduced along with the potential difference between the ends of the diode. In fact, the energy barrier or potential hill for holes in the

p-region has been reduced from qV_{bi} to

$$q(V_{bi} - V_A) \tag{3.4}$$

A similar reduction of the barrier height for the majority carrier electrons on the n-side is also illustrated in Fig. 3.3.

The reduced barrier for the diffusion of holes from the p-side to the n-side, for roughly the same carrier concentration gradient, yields an *increase in the hole diffusion current component* over the thermal equilibrium value. A large number of holes (p_p) have energies greater than the barrier height, $q(V_{bi} - V_A)$, as illustrated in Fig. 3.3, and therefore more holes can diffuse into the n-material, resulting in a larger diffusion current component. The hole drift current component remains the same as the thermal equilibrium value (small), since a change in barrier height has no effect on the number of holes (p_n) or their ability to drift down the potential hill.

The *net* flux of holes from left to right in the diagram represents a positive hole current since the number of holes able to *diffuse* is larger than those drifting back across the junction. Holes that are able to diffuse from the p-region into the n-region are called *injected holes* once they reach the n-region bulk, and are given the name "injected minority carriers" (more about this later).

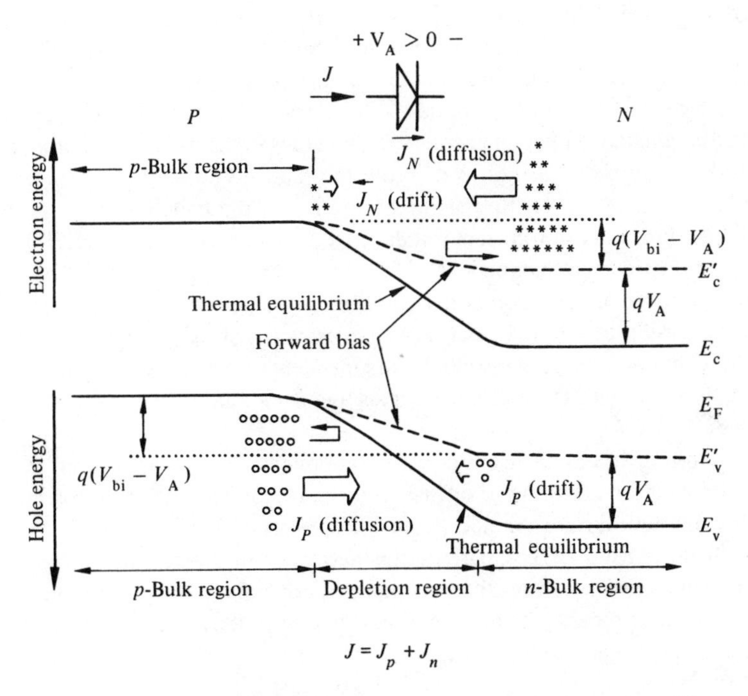

Fig. 3.3 Energy band diagram for forward bias and at thermal equilibrium.

Arguments for the electrons are similar to those for the holes. The barrier for electron diffusion from the n-region into the p-region is reduced, as illustrated in Fig. 3.3, and a large number of electrons are able to diffuse across the junction, causing a positive current. The majority carrier electrons reaching the p-material are called "injected minority carrier electrons" upon entering the p-bulk region.

Electrons from the p-region drift down the potential hill, but this component of current remains equal to its (small) thermal equilibrium value since the carrier supply is limited by thermal generation and not by the size of the potential hill.

The net effect of forward bias is a large increase in the diffusion current components while the drift components remain fixed near their thermal equilibrium values. Since the Fermi function distributes the carriers nearly exponentially with increasing energy, one should expect the number of carriers able to diffuse to increase exponentially with the reduction of the potential barriers. This being the case, the net forward bias current should increase exponentially with V_A, as will be shown later in this chapter.

3.2.2 Reverse Bias, $V_A < 0$

Reverse bias has $V_A < 0$, that is, V_A is a negative number and the polarity of the applied voltage is "positive" on the n-region contact. Figure 3.4 illustrates the effect on the energy band diagram as compared to thermal equilibrium. The n-bulk region* is shifted downward by $-qV_A$. The slope of the band edges in the depletion region has increased, reflecting the increase in electric field as discussed in Chapter 2 for reverse bias. Also note the increase in potential difference across the ends of the device, $q(V_{bi} - V_A)$, when $V_A < 0$. The applied voltage "adds" to the built-in potential.

The increase in barrier height encountered by holes in the p-region wanting to diffuse to the n-side is illustrated in Fig. 3.4. From the reduced number of holes (p_p) having sufficient energy to get over the potential barrier into the n-bulk region, it should be evident that the hole diffusion current is reduced to less than its thermal equilibrium value. However, the drift current component of holes from the n-region to the p-region down the potential hill remains at its thermal equilibrium value (small). Clearly the *net* hole current is from right to left and small in value. Actually, it is a small negative current. Because the drift component is essentially independent of barrier height, and because the carrier supply is limited, the current becomes a constant after several tenths of volts of reverse bias.

The electron drift and diffusion current components are also illustrated in Fig. 3.4. The n-region electrons wanting to diffuse to the p-region have a larger potential barrier than at thermal equilibrium. Hence, fewer electrons can make the trip and a smaller electron diffusion current, as compared to the thermal equilibrium value, is expected. The drift component remains at its thermal equilibrium value, limited by the supply of minority carrier electrons (n_p) in the p-region. Therefore, the *net* electron current is small in magnitude and negatively directed.

---

*This choice of the n-bulk region is arbitrary and the p-bulk could have been shifted upward by $(-qV_A)$, with the n-region fixed at the thermal equilibrium level.

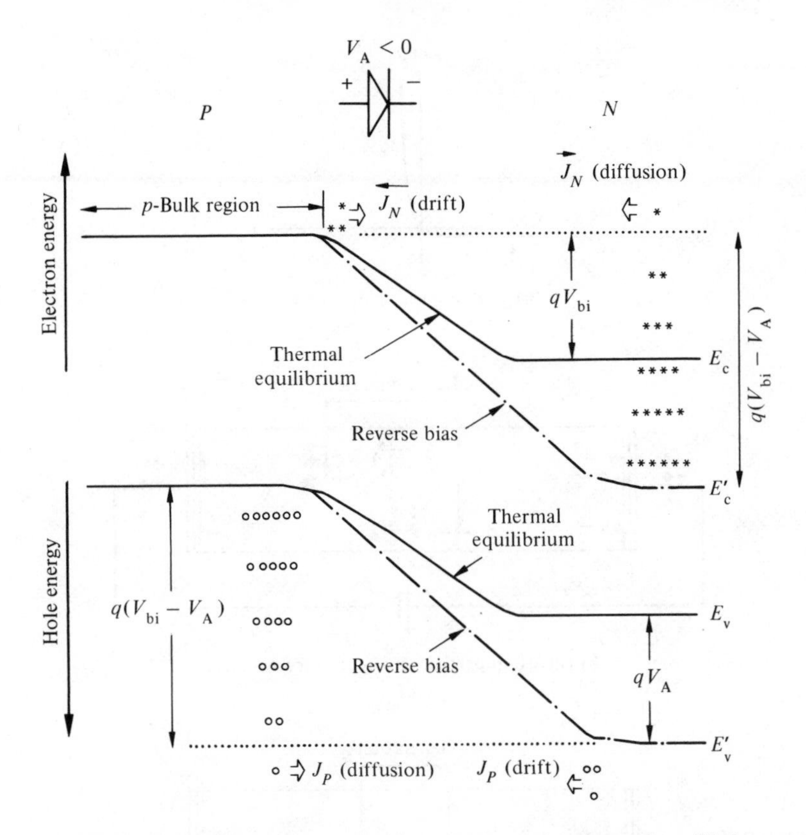

Fig. 3.4 Energy band diagram for reverse bias and at thermal equilibrium.

To summarize, the reverse current is small, negative, primarily limited by the supply of thermally generated minority carriers, and independent of V_A after a few tenths of a volt of reverse bias. Further increases in the amount of reverse bias have no effect on the minority carrier supply and reduce the diffusion current component to an insignificant value. The net result is a constant reverse current ($-I_0$). However, I_0 is very sensitive to changes in temperature, increasing with the supply of thermally generated minority carriers which are in turn proportional to n_i^2, the intrinsic carrier concentration, which increases exponentially with increasing temperature.

3.2.3 The Complete Circuit

Figure 3.5 is a qualitative representation of the current components in a diode under forward and reverse biasing. Note the exponential increase in the current for forward bias, $V_A > 0$, and the constant negative value of the reverse current for $V_A < 0$ in Fig. 3.5(a). The purpose of the metal contacts to the p and n material is illustrated in

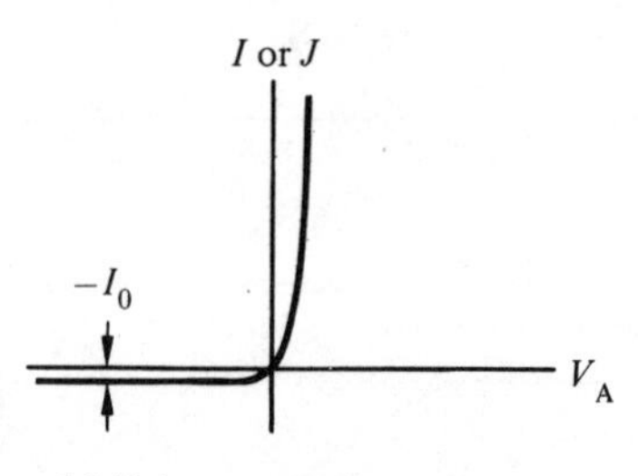

(a) Volt-ampere characteristic

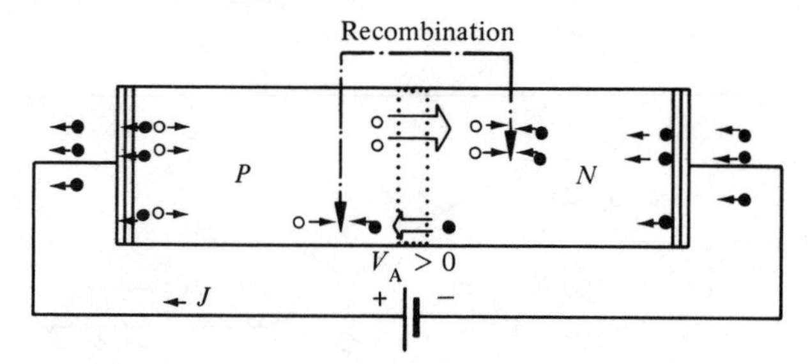

(b) Diffusion particle fluxes for forward bias

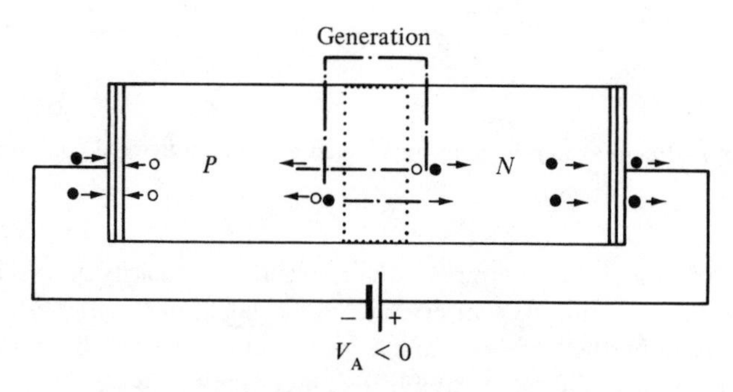

(c) Drift particle fluxes for reverse bias

Fig. 3.5 Diffusion and drift current components: (a) volt-ampere characteristic; (b) diffusion particle fluxes for forward bias; (c) drift particle fluxes for reverse bias.

Fig. 3.5(b). The metal–semiconductor contacts provide a mechanism for exchanging majority carriers in the semiconductor for electrons which can travel in the external circuit. Consider, in particular, the source of holes at the metal–p-material contact. To create a hole an electron must jump into the external circuit. The hole then moves toward the junction, eventually diffuses across the depletion region, and is injected into the

n-region as a minority carrier. As a minority carrier the hole has a very limited lifetime and soon recombines with a majority carrier electron. This recombination in turn calls for an electron to flow from the n-region contact as illustrated in Fig. 3.5(b). Note the continuous flow of carriers into and out of the semiconductor via the contacts.

The n-material has majority carrier electrons that diffuse across the depletion region, are injected into the p-material, and then recombine with majority carrier holes. Replacement holes are supplied at the metal–p-semiconductor contact by electrons exiting into the external circuit, thereby generating majority carrier holes. The continuity of current is completed at the metal–n-material contact, which provides the electrons for injection.

Figure 3.5(c), appropriate for reverse biases, illustrates the generation of drift current components near the depletion region edges. For example, the generation of minority carrier holes in the n-region drifts to the p-region, where they become majority carriers and eventually migrate to the metal–p-material contact. Electrons from the metal annihilate the excess holes causing an external circuit electron to flow. Simultaneously the thermally generated majority carrier electrons migrate to the metal–n-region contact and into the external circuit, completing the current loop.

The thermally generated electron–hole pairs near the p-side edge of the depletion region contribute to the reverse current by the electron's drifting across the depletion region to the n-material and as a majority carrier moves into the external circuit. Simultaneously the thermally generated hole near the p-side depletion region travels to the p-region contact and is exchanged for an external circuit electron, completing the current loop.

3.3 THE IDEAL DIODE EQUATION: DERIVATION GAME PLAN

This section develops a "game plan" for the quantitative solution of the basic semiconductor equations as applied to the abrupt p-n junction. The eventual goal is to derive a first-order relationship for the I versus V_A dependence of the p-n junction, known as the *ideal diode equation*. Subsequent sections carry out the mathematical details of the solution outlined in the present section.

The reader is again reminded of the explicit relationship between the p-n-junction diode and the myriad of other solid state devices, and that a thorough understanding of the diode is essential to understanding of other devices. Many of the mathematical derivations presented for the diode are used directly in modeling the bipolar and junction field effect transistors. Thus, the somewhat excessive time and effort allotted to the p-n junction will be amply repaid when analyzing other devices.

3.3.1 General Considerations

Chapter 3 of Volume I, Section 3.4 lists the "equations of state" for all semiconductor devices: the continuity equations, Poisson's equation, and the current flow equations. These equations are to be solved in each of the three regions of the p-n junction; the p-bulk region, the depletion region, and the n-bulk region. Chapter 2 of Volume II has discussed the electrostatic solution for the depletion region in detail. Several assumptions about the device are invoked to make the I–V_A solution tractable.

1. There are no external sources of generation, for example, no light.
2. The depletion approximation and the step junction are applicable.
3. The steady-state dc solution is desired; that is, all the d/dt terms are zero.
4. No generation or recombination takes place in the depletion region.
5. Low-level injection is maintained in the quasi-neutral (bulk) regions of the device.
6. The electric field for the minority carriers is zero in the bulk regions.
7. The bulk regions are uniformly doped; that is, N_A and N_D are constants.

With these assumptions the equations of state for the bulk n- and p-regions reduce to the following minority carrier equations.

n-type semiconductor.

$$0 = D_P \frac{d^2 \Delta p_n}{dx^2} - \frac{\Delta p_n}{\tau_p} \tag{3.5}$$

$$J_P \cong -qD_P \frac{d\,\Delta p_n}{dx} \tag{3.6}$$

$$p_n = p_{n0} + \Delta p_n(x) \tag{3.7}$$

p-type semiconductor.

$$0 = D_N \frac{d^2 \Delta n_p}{dx^2} - \frac{\Delta n_p}{\tau_n} \tag{3.8}$$

$$J_N \cong qD_N \frac{d\Delta n_p}{dx} \tag{3.9}$$

$$n_p = n_{p0} + \Delta n_p(x) \tag{3.10}$$

The reader should consult Section 3.4, Volume I, for the details concerning the origin of Eqs. (3.5) to (3.10).

The plan of attack is to first solve Eq. (3.5) for $\Delta p_n(x)$ in the n-bulk region. Because Eq. (3.5) is a second-order differential equation, two boundary conditions are needed, one at each end of the region. We will return to a consideration of the boundary conditions later. Once $\Delta p_n(x)$ is obtained, then Eq. (3.6) is used to compute $J_P(x)$. Since the device has only two terminals, the total current through the diode must be a constant at each point:

$$J = \text{constant} = J_N(x) + J_P(x) \tag{3.11}$$

Therefore, if the minority carrier current density $J_P(x)$ is known in the n-bulk region, the majority carrier current density $J_N(x)$ is also known from Eq. (3.11) as

$$J_N(x) = J - J_P(x) \tag{3.12}$$

Arguments for the p-bulk region and the minority carrier electrons are complementary to those of the minority carrier holes in the n-bulk region. By complementary we mean the exchange of p for n and n for p. For example, the complement of Eq. (3.5) is Eq. (3.8). A solution of Eq. (3.8) yields $\Delta n_p(x)$ and the use of Eq. (3.9) results in $J_N(x)$, the minority carrier current in the bulk p-region. The majority carrier current is obtained from Eq. (3.11) as

$$J_P(x) = J - J_N(x) \tag{3.13}$$

The perceptive student might ask, "You have outlined a plan of attack for the bulk regions, but what about the current in the depletion region?" The depletion region was assumed to have no generation or recombination. Therefore, the current through it is a constant—what goes in must come out. No carriers are added to or subtracted from the carrier flux. Figure 3.6 illustrates this point. Note that if the minority carrier diffusion current is known at the *edges* of the depletion region, then it is known throughout the depletion region. Specifically,

$$J_{P|\text{depl}} = J_P(x_n) = -qD_P \left.\frac{dp_n}{dx}\right|_{x=x_n} \tag{3.14}$$

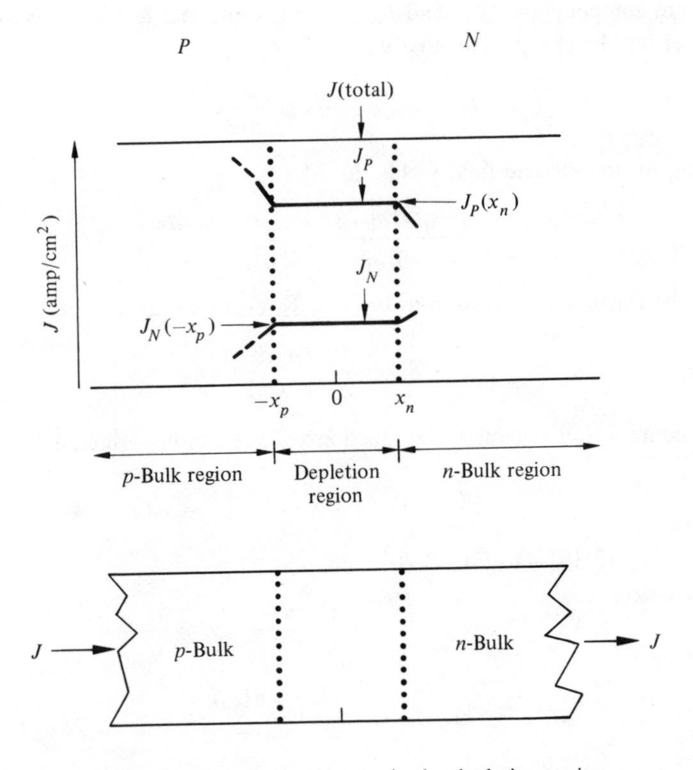

Fig. 3.6 Current components in the depletion region.

$$J_{N|\text{depl}} = J_N(-x_p) = qD_N \frac{dn_p}{dx}\bigg|_{x=-x_p} \tag{3.15}$$

Finally, as illustrated in Fig. 3.6, the total current is simply the sum of Eq. (3.14) and Eq. (3.15).

$$J = J_P(x_n) + J_N(-x_p) \tag{3.16}$$

The reader may well ask, "Where does the applied voltage (V_A) get into the act? You started out to derive a J versus V_A relationship, but V_A has yet to be even mentioned!" As it turns out, V_A enters via the boundary conditions on $p_n(x)$ and $n_p(x)$ at the edges of the depletion region in the solution to the minority carrier diffusion equations.

3.3.2 Boundary Conditions at x_n and $-x_p$

In Chapter 2 when V_A was applied across the terminals of the diode, the junction potential (V_j) was equated to $V_{bi} - V_A$ under the assumption that the electric field was essentially zero in the bulk regions. The assumption of "low-level injection" suggests that in the depletion region the additional currents due to the applied voltage are also small. *In the depletion region* $\mathscr{E} \neq 0$ and the electron current is the difference between the large current components, $J_{N|\text{drift}}$ and $J_{N|\text{diff.}}$. We assume that $\mathscr{E}$ and n have not changed much under low-level injection, that is,

$$J_N = q\mu_n n\mathscr{E} + qD_N \frac{dn}{dx} \cong 0 \tag{3.17}$$

and solving for the electric field yields Eq. (3.18),

$$\mathscr{E} = \frac{-qD_N\, dn/dx}{q\mu_n n} = \frac{-D_N}{\mu_n} \frac{dn/dx}{n} \tag{3.18}$$

Applying the Einstein relationship to Eq. (3.18) results in Eq. (3.19),

$$\mathscr{E} = \frac{-kT}{q} \frac{dn/dx}{n} \tag{3.19}$$

From the definition of potential, the junction voltage can be written as

$$V_j = V_{bi} - V_A = -\int_{-x_p}^{x_n} \mathscr{E}\, dx \tag{3.20}$$

Substituting Eq. (3.18) into Eq. (3.20) yields

$$V_j = V_{bi} - V_A = -\int_{-x_p}^{x_n} -\frac{kT}{q} \frac{dn/dx}{n}\, dx = \frac{kT}{q} \ln n \bigg|_{n(-x_p)}^{n(x_n)} \tag{3.21}$$

$$V_{bi} - V_A = \frac{kT}{q} \ln \frac{n(x_n)}{n(-x_p)} \tag{3.22}$$

Remember that for the majority carriers in the bulk region $\Delta n_n \ll n_{n0}$ for the "low-level injection" requirement to be valid. Therefore $n(x_n) = n_{n0}$. Inverting Eq. (3.22) by cross multiplying and raising both sides to the exponential, we get the electron concentration ratio:

$$\frac{n(x_n)}{n(-x_p)} = e^{q(V_{bi}-V_A)/kT} = e^{[qV_{bi}/kT]}e^{[-qV_A/kT]} \tag{3.23}$$

Solving for $n(-x_p)$ results in Eq. (3.24)

$$n(-x_p) = n(x_n)e^{[-qV_{bi}/kT]}e^{[qV_A/kT]} \tag{3.24}$$

From Eq. (2.14), for thermal equilibrium,

$$V_{bi} = \frac{kT}{q} \ln\left[\frac{n_{n0}p_{p0}}{n_i^2}\right] \tag{3.25}$$

which can be inverted to obtain Eq. (3.26),

$$e^{-qV_{bi}/kT} = \frac{n_i^2}{n_{n0}p_{p0}} \tag{3.26}$$

Combining Eq. (3.24) and Eq. (3.26) yields

$$n(-x_p) = n(x_n)\frac{n_i^2}{n_{n0}p_{p0}}e^{qV_A/kT} = \frac{n_i^2}{p_{p0}}e^{qV_A/kT} \tag{3.27}$$

remembering that $n(x_n) = n_{n0}$. Since $n_i^2/p_{p0} = n_{p0}$ one concludes that

$$n(-x_p) = n_{p0}\, e^{qV_A/kT} \tag{3.28}$$

and the excess electron concentration at $-x_p$ is

$$[\![\Delta n(-x_p) = n(-x_p) - n_{p0} = n_{p0}(e^{qV_A/kT} - 1)]\!] \tag{3.29}$$

Note that Eqs. (3.28) and (3.29) reduce to their thermal equilibrium values when $V_A = 0$.

Complementary arguments can be used for the hole concentration at the edges of the depletion regions and result in Eqs. (3.30) and (3.31),

$$p_n(x_n) = p_{n0}e^{qV_A/kT} \tag{3.30}$$

$$[\![\Delta p_n(x_n) = p_{n0}(e^{qV_A/kT} - 1)]\!] \tag{3.31}$$

Long Base Diode. The final boundary conditions on the excess carrier concentration in the p- and n-bulk regions are obtained by assuming that the bulk regions are very long, infinite, in length. Since the excess minority carriers have a finite lifetime (τ_p and τ_n), they cannot survive forever without recombining; consequently,

$$\Delta n_p(-\infty) = 0 \tag{3.32}$$

and

$$\Delta p_n(+\infty) = 0 \tag{3.33}$$

Game Plan Summary

1. Solve the minority carrier continuity equations in the bulk regions for $\Delta p_n(x)$ and $\Delta n_p(x)$ or $p_n(x)$ and $n_p(x)$.

2. Apply two boundary conditions to each solution to determine $\Delta p_n(x)$ and $\Delta n_p(x)$ in terms of the applied voltage V_A.

3. Determine the currents $J_P(x_n)$ and $J_N(-x_p)$ from the *slope* of $\Delta n_p(x)$ and $\Delta p_n(x)$ at $-x_p$ and x_n respectively, using Eqs. (3.14) and (3.15).

4. The total current is then the sum of the currents at the edges of the depletion region; that is,

$$J = J_P(x_n) + J_N(-x_p) \tag{3.34}$$

3.4 THE IDEAL DIODE EQUATION: DERIVATION

To solve the minority carrier diffusion equations, Eqs. (3.5) and (3.8), we begin by selecting the special x' and x'' coordinate systems as illustrated in Fig. 3.7. The translation of the working-coordinate system simplifies the analytical form of the solutions and avoids unnecessary complications in the analysis.

3.4.1 *n*-Bulk Region, $x \geqq x_n$ or $x' \geqq 0'$

Equation (3.35) is Eq. (3.5) rewritten in terms of the new variable x' whose origin is at $x = x_n$.

$$D_P \frac{d^2\Delta p_n(x')}{dx'^2} - \frac{\Delta p_n(x')}{\tau_p} = 0 \tag{3.35}$$

Dividing by D_P and taking the second term across the equal sign results in Eq. (3.36), where L_P is defined as the *minority carrier diffusion length for holes*:

$$\frac{d^2\Delta p_n(x')}{dx'^2} = \frac{\Delta p_n(x')}{D_P\tau_p} = \frac{\Delta p_n(x')}{L_P^2} \tag{3.36}$$

$$L_P = \sqrt{D_P\tau_p} \quad \text{cm} \tag{3.37}$$

Equation (3.36) is a very common differential equation and can be solved directly or by Laplace transforms. The solution is of the form

$$\Delta p_n(x') = A_1 e^{x'/L_P} + A_2 e^{-x'/L_P} \tag{3.38}$$

where two boundary conditions are needed to evaluate the constants A_1 and A_2. From Eq. (3.33) of the previous section, the boundary condition at infinity requires

$$\Delta p_n(\infty) = A_1 e^{\infty} + A_2 e^{-\infty} = A_1 e^{\infty} + 0 = 0 \tag{3.39}$$

Equation (3.39) can be satisfied only if $A_1 = 0$. The second boundary condition, Eq. (3.31), requires

$$\Delta p_n(x_n) = \Delta p_n(0') = p_{n0}(e^{qV_A/kT} - 1) = A_2 e^{-0'/L_P} = A_2 \tag{3.40}$$

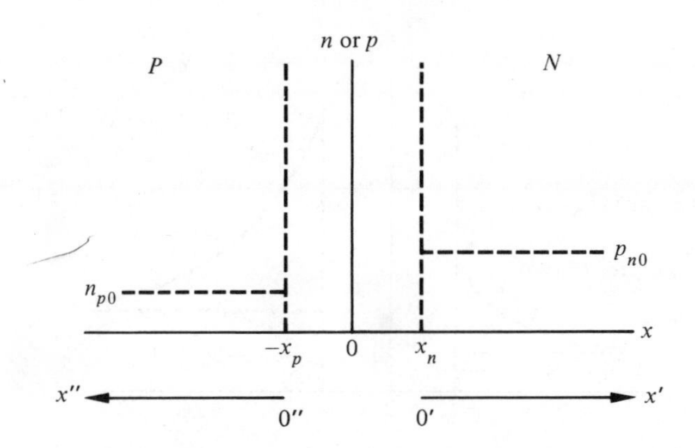

Fig. 3.7 Axis selection for bulk regions.

Therefore the solution for $\Delta p_n(x)$ is

$$\Delta p_n(x') = p_{n0}(e^{qV_A/kT} - 1)e^{-x'/L_P} \tag{3.41}$$

or

$$p_n(x') = p_{n0} + p_{n0}(e^{qV_A/kT} - 1)e^{-x'/L_P} \tag{3.42}$$

Equation (3.42) is plotted on the right-hand side of Fig. 3.8 for $V_A > 0$ and $V_A < 0$.

Following the game plan, we next obtain the hole current by applying Eq. (3.14) to Eq. (3.41):

$$J_P(x') = -qD_P\frac{d\Delta p_n}{dx'} = -qD_P p_{n0}(e^{qV_A/kT} - 1)\left(\frac{-1}{L_P}\right)e^{-x'/L_P} \tag{3.43a}$$

$$J_P(x') = \frac{qD_P}{L_P}p_{n0}(e^{qV_A/kT} - 1)e^{-x'/L_P} \tag{3.43b}$$

Evaluating at $x = x_n$ yields the hole current in the depletion region,

$$J_{P|\text{depl}} = J_P(x_n) = J_P(0') = q\frac{D_P}{L_P}p_{n0}(e^{qV_A/kT} - 1) \tag{3.44}$$

Figure 3.9 illustrates the hole current in the depletion and n-bulk regions.

3.4.2 *p*-Bulk Region, $x \leqq -x_p$ or $x'' \geqq 0$

The complementary p-bulk solution to Eq. (3.8) can be established directly or derived step-by-step in a manner similar to the n-bulk derivation. Simply taking the complement of Eqs. (3.41) and (3.42) yields Eqs. (3.45) and (3.46),

$$\Delta n_p(x'') = n_{p0}(e^{qV_A/kT} - 1)e^{-x''/L_N} \tag{3.45}$$

$$n_p(x'') = n_{p0} + n_{p0}(e^{qV_A/kT} - 1)e^{-x''/L_N} \tag{3.46}$$

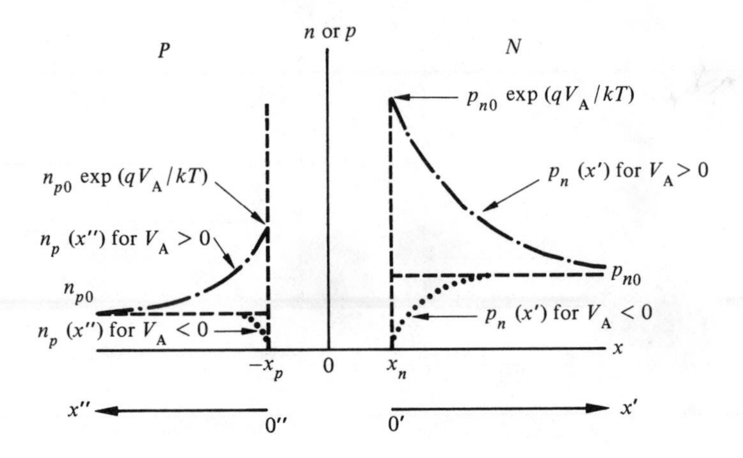

Fig. 3.8 Minority carrier concentrations in the bulk regions.

The *minority carrier diffusion length for electrons* has been defined as

$$L_N = \sqrt{D_N \tau_n} \qquad \text{cm} \tag{3.47}$$

The left-hand side of Fig. 3.8 illustrates these results for forward and reverse bias. By applying the complement to Eqs. (3.43) and (3.44), the electron currents $J_N(x'')$ and

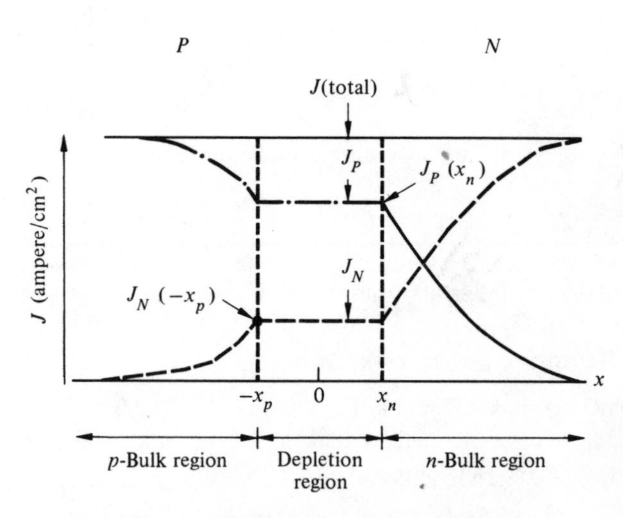

Fig. 3.9 Current density in the forward biased case.

$J_N(0'') = J_N(-x_p)$ are readily determined to be

$$J_N(x'') = q\frac{D_N}{L_N}n_{p0}(e^{qV_A/kT} - 1)e^{-x''/L_N} \tag{3.48}$$

$$J_N(-x_p) = q\frac{D_N}{L_N}n_{p0}(e^{qV_A/kT} - 1) \tag{3.49}$$

Figure 3.9 illustrates the electron current components throughout the diode.

3.4.3 Ideal Diode Equation

The total current is obtained by adding Eq. (3.44) and Eq. (3.49) as outlined in Step 4 of the "game plan," that is Eq. (3.34). The result is called the ideal diode equation or sometimes the Shockley diode equation,

$$J = q\left[\frac{D_N}{L_N}n_{p0} + \frac{D_P}{L_P}p_{n0}\right](e^{qV_A/kT} - 1) \tag{3.50}$$

Multiplying by the area of the junction (A) yields

$$I = I_0(e^{qV_A/kT} - 1) \tag{3.51}$$

where the reverse saturation current or reverse leakage current has been defined as

$$I_0 = qA\left[\frac{D_N}{L_N}n_{p0} + \frac{D_P}{L_P}p_{n0}\right] \tag{3.52}$$

3.5 INTERPRETATION OF RESULTS

3.5.1 V–I Relationship

Further examination of the quantitative solution for the ideal diode equation leads to a deeper insight into the operation of the p-n junction. Figure 3.10(a) is a linear plot of Eq. (3.51) illustrating the exponential increase in current with forward bias and the small, nearly constant, reverse bias current. Examination of the exponential term in Eq. (3.51) at room temperature, where $q/kT = 38.46\ \text{V}^{-1}$, indicates that a forward bias of 0.10 volts makes the exponential nearly 50 times that of the "one," and therefore, the current is exponentially dependent on the applied voltage. Similarly, with $V_A = -0.10$ volts, the exponential is about 1/50 and very small compared to the "-1," leaving the current $I = -I_0$ independent of the reverse voltage. The reverse current no longer changes, that is, becomes saturated; hence its name "the reverse saturation current."

As evidenced by Eq. (3.52) and as visualized in Fig. 3.4, the reverse saturation current is determined primarily by the minority carriers n_{po} thermally generated in the p-bulk region and p_{n0} generated in the n-bulk region. Equation (3.52) can be written in

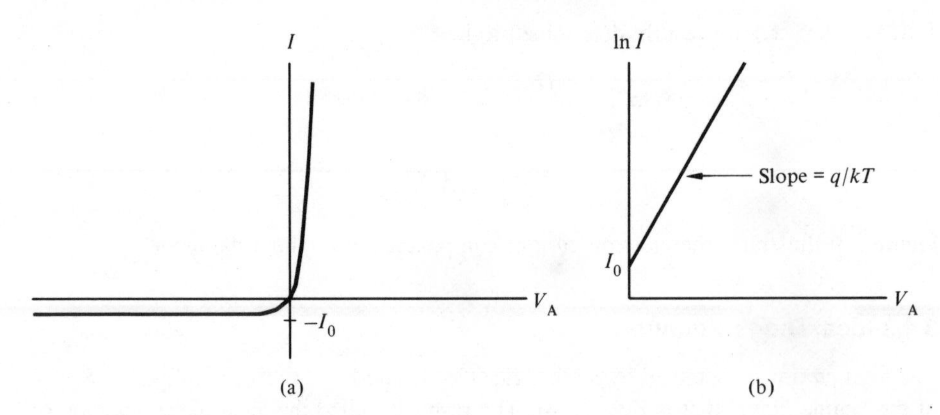

Fig. 3.10 (a) *p-n* junction volt-ampere characteristics; linear and (b) semilogarithmic.

terms of the doping densities N_A and N_D:

$$I_0 = qA\left[\frac{D_N}{L_N}n_{p0} + \frac{D_P}{L_P}p_{n0}\right] = qA\left[\frac{D_N}{L_N}\frac{n_i^2}{N_A} + \frac{D_P}{L_P}\frac{n_i^2}{N_D}\right] \tag{3.53}$$

It is important to note that n_i increases exponentially with temperature and its room-temperature value is very dependent on the bandgap E_G of the material. A larger E_G value results in a smaller value of n_i. Also note that the lighter doped side of the p-n junction will produce a larger number of minority carriers, and thus the larger current component. For example, a p^+-n junction has $N_A \gg N_D$ and, from Eq. (3.53),

$$I_0 \cong qA\left[\frac{D_P}{L_P}p_{n0}\right] = qA\left[\frac{D_P}{L_P}\frac{n_i^2}{N_D}\right] \tag{3.54}$$

Similarly for p-n^+ junction,

$$I_0 \cong qA\left[\frac{D_N}{L_N}n_{p0}\right] = qA\left[\frac{D_N}{L_N}\frac{n_i^2}{N_A}\right] \tag{3.55}$$

Most diodes and many of the p-n junctions that occur in other devices are of the p^+-n or n^+-p type, and we shall make frequent use of these asymmetrical junctions in future analyses.

3.5.2 Current Components

The forward biased electron and hole currents illustrated in Fig. 3.9 are for the assumed case of $N_A > N_D$, that is, $p_{n0} > n_{p0}$. Inspection of Eqs. (3.43) and (3.48) indicates that the hole current injected into the n-region is larger than the electron current injected into the p-region when $N_A > N_D$. This is a very important point, a "V. I. P." The p^+-n junction in particular obtains most of its current from the holes injected into the n-region. Similarly the p-n^+ junction current is composed mostly of electrons injected into the p-region.

3.5.3 Carrier Concentrations

The minority carrier concentrations are plotted for forward bias in Fig. 3.11. Cross-hatching emphasizes the injected excess-carrier concentrations. From Eqs. (3.42) and (3.46), the rates of exponential decay of the injected minority carriers, as they recombine with majority carriers, are controlled by the minority carrier diffusion lengths L_P and L_N. Figure 3.11 illustrates L_P and L_N as the distance into the bulk region where the excess carriers have decreased to 37% of their value at the depletion-region edge. Since L_P and L_N are proportional to $\tau_p^{1/2}$ and $\tau_n^{1/2}$, the greater the recombination (the smaller τ_p and τ_n), the shorter the diffusion lengths. Another interpretation is that on the average a minority carrier will diffuse into the bulk region one diffusion length before recombining.

The reverse bias carrier concentrations are plotted in Fig. 3.12. In this case the "excess" carrier concentrations are negative. On the average, the minority carriers

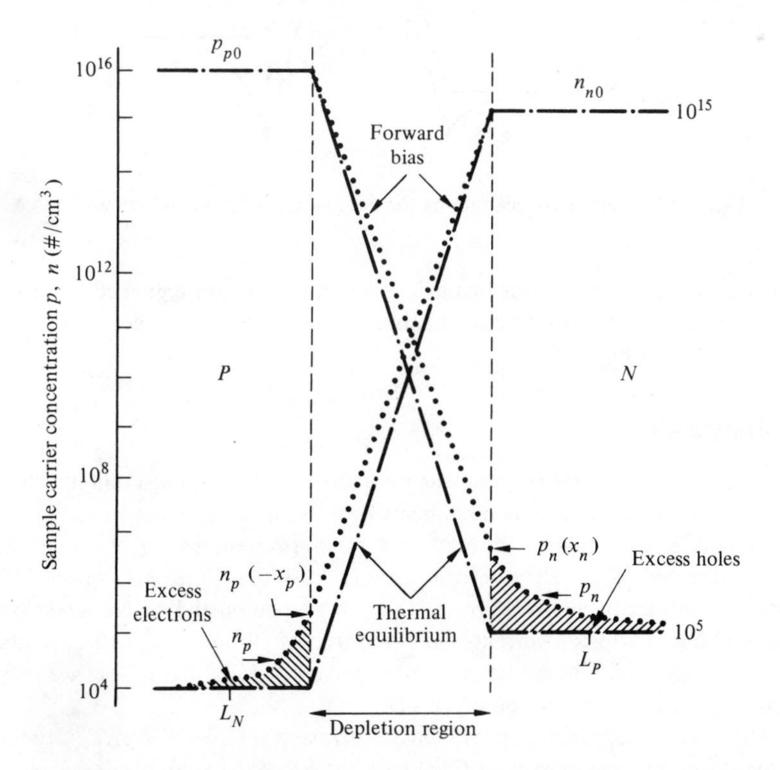

Fig. 3.11 Carrier concentrations at thermal equilibrium and at forward bias.

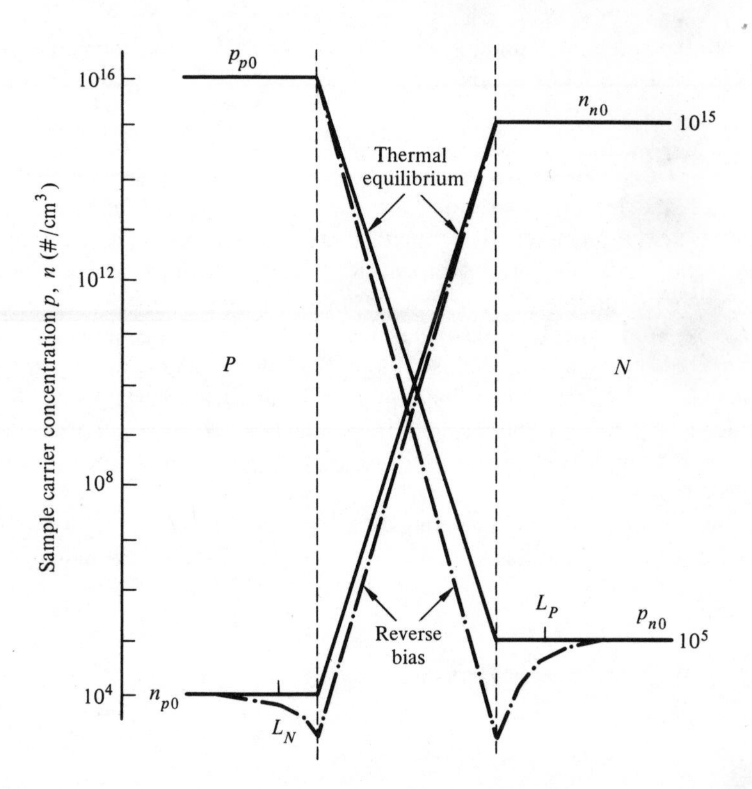

Fig. 3.12 Carrier concentrations for thermal equilibrium and reverse bias.

thermally generated within one diffusion length of the depletion-region edge are those that are encouraged to drift down the potential hill and contribute to I_0. These are the minority carriers depicted in Fig. 3.1.

3.6 SUMMARY

The thermal equilibrium, energy band diagram served as the starting point for carrier flow by diffusion and drift. The two components of hole current counteract each other for a net hole flux of zero. Similarly, the electron diffusion current was equal in magnitude but oppositely directed to the electron drift current, resulting in no net current. The drift currents were primarily controlled by the thermal generation and the carriers' flow down the potential hill. Diffusion currents are comprised of carriers with sufficient energy to surmount the potential hill and have a large concentration of carriers on one side of the depletion region compared to the other side.

The forward biased junction permitted the diffusion currents to increase exponentially while the drift components remain fixed near their thermal equilibrium values. Excess holes are injected into the n-region and excess electrons are injected into the p-region.

Reverse bias was characterized by the diffusion currents reaching a very small value while the drift current remained nearly fixed and dominated the conduction after several tenths of reverse voltage.

The ideal diode equation was based on the depletion approximation and low-level injection. With the currents constant within W, we used the minority carrier concentrations at the edges of the depletion width to evaluate the currents. The junction potential controlled the minority carrier concentrations at the edges of W.

PROBLEMS

3.1 A silicon step junction has $N_A = 5 \times 10^{15}/\text{cm}^3$ and $N_D = 10^{15}/\text{cm}^3$ (see Problem 2.1) with $A = 10^{-4}\text{cm}^2$ and $\tau_p = 0.4\ \mu\text{sec}$, $\tau_n = 0.1\ \mu\text{sec}$. Calculate

(a) Leakage current due to holes.

(b) Leakage current due to electrons.

(c) I_0

(d) If $V_A = V_{bi}/2$ calculate:

(i) hole concentration at x_n; excess hole concentration at x_n.

(ii) hole concentration at $x' = L_P/2$.

(iii) electron concentration at $-x_p$; excess electron concentration at $-x_p$.

(iv) electron concentration at $x'' = L_N/2$.

(e) If $V_A = -V_{bi}/2$ calculate:

(i) hole concentration at x_n and $x' = L_p/2$.

(ii) electron concentration at $-x_p$ and $x'' = L_N/2$.

(f) Calculate the total excess hole charge for:

(i) part (d).

(ii) part (e).

(g) At what value of applied voltage will the "low-level injection" assumption first be violated? Use a one-to-ten ratio as the criterion.

3.2 A p-n abrupt junction with $N_A = 10^{17}/\text{cm}^3$ and $N_D = 5 \times 10^{15}/\text{cm}^3$ has $\tau_p = 0.1$ microseconds and $\tau_n = 0.01\ \mu\text{sec}$. $A = 10^{-4}\text{cm}^2$.

(a) Calculate the leakage current due to holes.

(b) Calculate the leakage current due to electrons.

(c) Calculate the total leakage current.

(d) If $V_A = V_{bi}/2$, calculate the injected minority carrier currents. What is the excess carrier concentration at 0 and 1 micron into the bulk region?

(e) If $V_A = -V_{bi}/2$, calculate the minority carrier concentrations at the edge of the depletion region.

(f) At what value of applied voltage will the "low-level injection" assumption fail? [Use 1/10 as the limit of the carrier ratio.]

3.3 Two p^+-n Ge diodes maintained at room temperature are identical except that $N_{D1} = 10^{15}/\text{cm}^3$ while $N_{D2} = 10^{16}/\text{cm}^3$. Sketch and compare the $I-V$ characteristics of the two diodes.

3.4 A Si step-junction diode with a cross-sectional area (A) of 10^{-4}cm^2 has a doping of $N_A = 10^{17}/\text{cm}^3$ on the p-side and $N_D = 10^{15}/\text{cm}^3$ on the n-side. Let

$$\mu_n = 801 \text{ cm}^2/\text{V-sec and } \tau_n = 0.1\ \mu\text{sec on the } p\text{-side;}$$

$$\mu_p = 477 \text{ cm}^2/\text{V-sec and } \tau_p = 1\ \mu\text{sec on the } n\text{-side.}$$

(a) Assuming that the diode characteristics are described by the ideal diode equation, compute the current (I) through the diode at room temperature for (i) $V_A = -50\text{V}$, (ii) $V_A = -0.1\text{V}$ and (iii) $V_A = 0.2\text{V}$.

(b) Assuming the μ's and τ's do not change with temperature, repeat parts (a) for an operational temperature of $T = 500\ °\text{K}$.

(c) Summarize in words what has been exhibited mathematically in this problem.

3.5 Consider the situation shown in Fig. P3.5 which is an idealization of the actual situation inside a diode under reverse bias conditions:

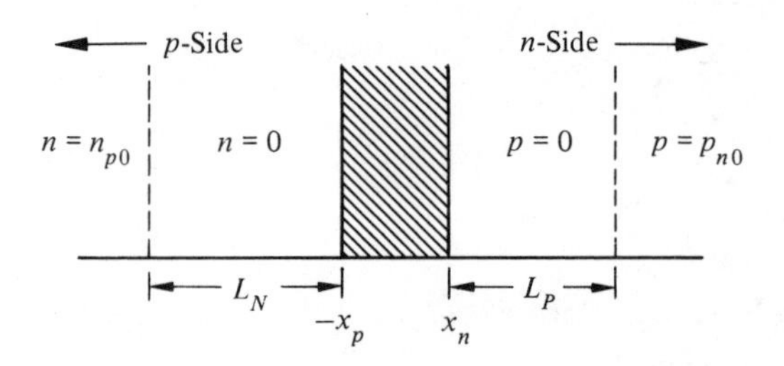

Figure P3.5

(a) Derive an expression for the number of electrons generated per second in the $n = 0$ region shown above. (*Hint:* $\partial n/\partial t|_{R-G} = -\Delta n/\tau_n$.)

(b) If all of the electrons generated per second are swept out of the $n = 0$ region across the depletion region, what will be the associated current flow?

(c) Repeat (a) and (b) for holes in the $p = 0$ region.

(d) Add the I_N and I_P expressions obtained in (b) and (c).

(e) Show that the I obtained in (d) is nothing more than $-I_0$, the current flowing in an ideal diode under reverse bias conditions where $-V_A >$ few kT/q.

(f) Explain the significance of this problem.

3.6 In Fig. P3.6:

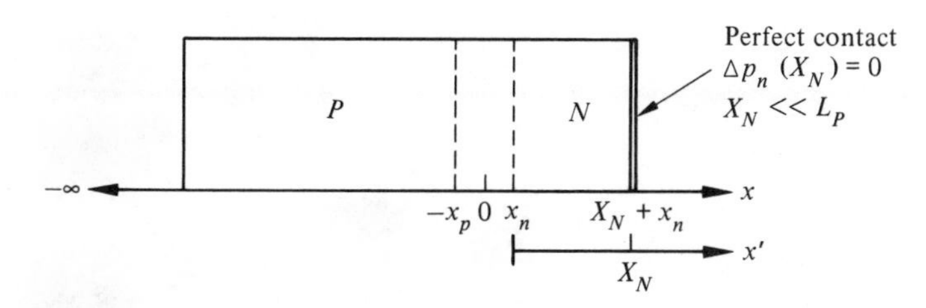

Figure P3.6

(a) Derive an equation for $\Delta p_n(x')$ and plot $p_n(x)$ if forward biased.

(b) Derive J_P in the n-region and plot it for forward bias.

(c) Use previous results for J_N in the p-region and obtain an equation for J.

(d) Plot parts (a) and (b) for reverse bias.

3.7 If the step junction is forward biased:

(a) Derive an equation for the total excess minority carrier charge in the n-region, Q_p.

(b) Take the complement of part (a) and obtain Q_n for the p-region.

(c) Compare Q_p and $I_{p|\text{diff}}$ for holes and obtain an equation for I_P at x_n. Similarly for I_n at $-x_p$. Obtain a new ideal diode equation in terms of Q_n and Q_p.

(d) Make a statement about recombination and current as a result of part (c). *Hint:* τ is the average time to recombine.

4 / Deviations from the Ideal Diode

The first three chapters of this volume have developed the qualitative and quantitative descriptions of the ideal *p-n* junction diode. The ideal diode equation, as derived, accurately describes certain real devices over many decades of current and over a range of applied voltages. However, several conditions of applied voltage and/or temperature exist where the ideal diode fails to adequately represent physical devices. When reverse biased, the current can become more negative than $-I_0$ as a result of generation of carriers in the depletion region, and at even larger reverse voltages due to junction breakdown. Breakdown is due to one of two phenomena: avalanching or the Zener process. Forward bias deviations from the ideal occur at very small currents due to recombination in the depletion region, and at very large currents due to two effects: high-level carrier injection and ohmic voltage drops in the bulk regions and contacts.

4.1 REVERSE BIAS DEVIATIONS FROM IDEAL

The deviations from ideal for reverse bias are illustrated in Fig. 4.1. At large values of reverse bias the magnitude of the current increases dramatically. The voltage where the current tends toward $-\infty$ is called the *breakdown voltage* (V_{BR}). Should the diode operate in this region it is not destroyed, as would be the case for a capacitor, provided the maximum junction temperature is not exceeded. In a number of applications the diode is operated in the breakdown region and behaves like a nearly fixed voltage source over a large range of currents; hence the name "reference diode" or "Zener diode." Operated as a voltage reference in breakdown, the diode replaces a large expensive battery in many voltage regulator type circuits.

The mechanisms responsible for V_{BR} are avalanching and/or the Zener process. Avalanching is more common and will therefore be discussed first.

4.1.1 Avalanche Breakdown

As a starting point for discussing avalanche phenomena, consider the electric field in the depletion region and the minority carriers drifting down the potential hill, giving rise to the current $-I_0$. Chapter 2 discussed the effect of larger reverse voltages and their resultant larger electric fields in the depletion region, with the maximum electric field occurring at the metallurgical junction. Remember that the larger the electric field, the larger the drift velocity of the carriers. At a critical value of electric field ($\mathscr{E}_{CR}$), the carriers, on the average, accelerate to a large enough energy that when they collide with the crystal lattice, they "free" an electron–hole pair. The colliding carrier imparts enough energy to the crystal that an electron in the valence band is excited to the conduction band, leaving a hole in the valence band. All three carriers are now "free" to be accelerated by the electric field and participate in additional carrier creating collisions; that is, an avalanche takes place. A single carrier, either an electron or a hole, creates two carriers, which in turn create additional electron–hole pairs. Figure 4.2(a) illustrates the process.

Note that the holes entering the depletion region edge are multiplied by the avalanching phenomenon until they approach the p-bulk region and the electric field falls below the critical value. The ratio of the hole current leaving the depletion region to the hole current entering is called the multiplication factor for holes with a similar definition for electrons. The total multiplication factor is given by Eq. (4.1):

$$M = \frac{I_{\text{out}}}{I_{\text{in}}} = \frac{|I|}{I_0} \tag{4.1}$$

An expression for the multiplication factor in terms of the breakdown voltage will not be derived here, but the result is

$$M = \frac{1}{1 - \left[\frac{|V_A|}{V_{BR}}\right]^m} \tag{4.2}$$

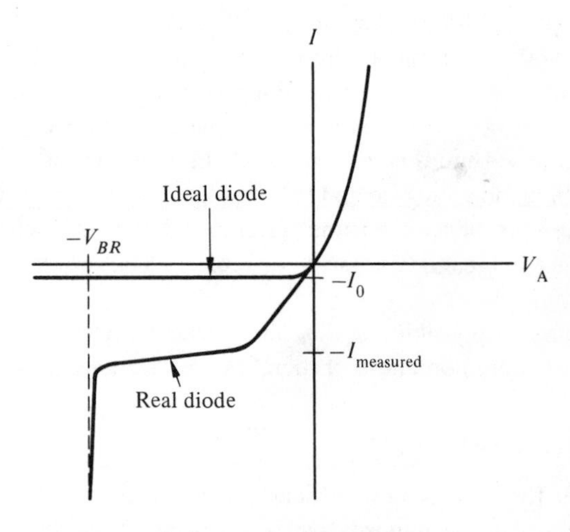

Fig. 4.1 Reverse bias deviations from ideal.

where m has the range of $3 \leq m \leq 6$ depending on the semiconductor employed. Figure 4.1 and Fig. 4.2(b) illustrate that, near the knee of the $V-I$ curve, significant multiplication occurs before the breakdown voltage is reached.

The effect of doping on the breakdown voltage will now be considered. In Chapter 2 the maximum electric field occurred at $x = 0$. Combining Eqs. (2.47) and (2.45), the maximum negative electric field is

$$\mathscr{E}(0) = \frac{-qN_D}{K_S\varepsilon_0}(x_n - 0) = -\left[\frac{2q}{K_S\varepsilon_0}(V_{bi} - V_A)\frac{N_A N_D}{N_A + N_D}\right]^{1/2} \tag{4.3}$$

The critical value of electric field ($\mathscr{E}_{CR}$) is a physical constant for a given semiconductor and is independent of doping. Replacing V_A by $-V_{BR}$ in Eq. (4.3) and squaring yields

$$\mathscr{E}_{CR}^2 = \frac{2q}{K_S\varepsilon_0}(V_{bi} + V_{BR})\frac{N_A N_D}{N_A + N_D} = \text{constant} \tag{4.4}$$

Typically $V_{BR} \gg V_{bi}$, and one therefore concludes,

$$V_{bi} + V_{BR} \cong V_{BR} \propto \frac{N_A + N_D}{N_A N_D} \tag{4.5}$$

An examination of Eq. (4.5) indicates that any increase in doping, either n or p, results in a decrease in V_{BR}. For the p^+-n junction,

$$V_{BR} \propto \frac{1}{N_D} \tag{4.6}$$

and for the p-n^+ junction

$$V_{BR} \propto \frac{1}{N_A} \tag{4.7}$$

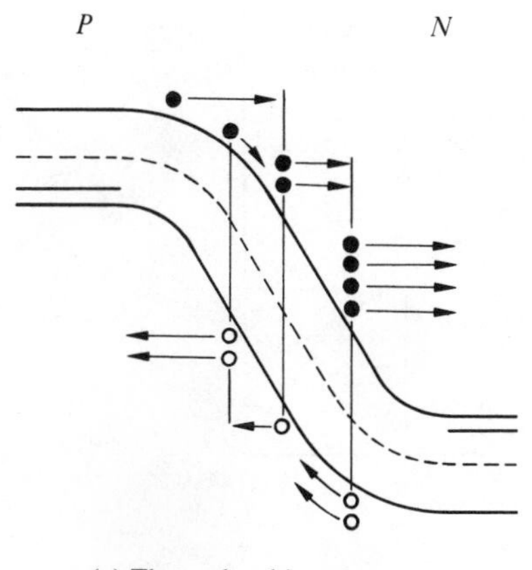

(a) The avalanching process

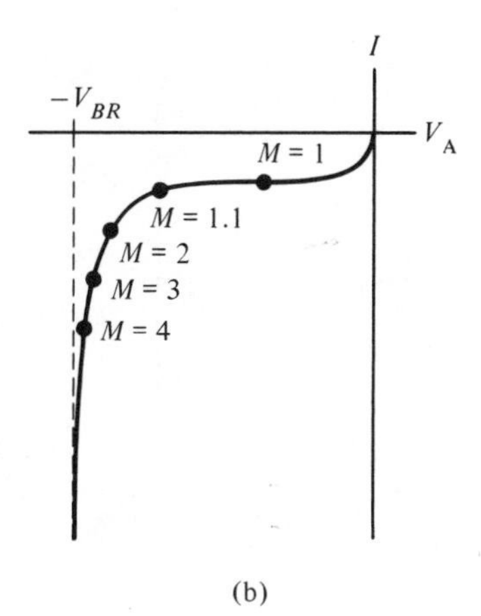

(b)

Fig. 4.2 The avalanching process: (a) multiplication; (b) breakdown voltage.

The breakdown voltage, due to avalanching, is primarily controlled by the doping of the lightly doped bulk region. Figure 4.3 illustrates the above cases more explicitly. For silicon any V_{BR} greater than 5 or 6 volts is a result of avalanching.

4.1.2 Zener Breakdown

Zener breakdown occurs in *p-n* junctions that are heavily doped on both sides of the metallurgical junction. Typically Si diodes with V_{BR} of 4 volts or less meet the criteria for *tunneling*, another name for the Zener process. Figure 4.4 illustrates the basic idea of an electron tunneling through a potential barrier. In classical physics a particle must have an energy greater than the barrier to appear on the other side. However, quantum mechanically, if the barrier is very thin, $d < 100$ Å, the carrier may tunnel through the barrier. The two basic requirements for tunneling are:

1. A thin potential barrier, that is, the smaller d is, the larger the probability of tunneling.
2. A large number of electrons available to tunnel on one side of the barrier and a large number of empty states, at the same energy level, into which to tunnel on the other side of the barrier.

A reverse-biased *p-n* junction diode is illustrated in Fig. 4.5. The criteria for tunneling, as applied to the diode, are met if W is small. The depletion width is small if both the *p*- and *n*-sides have large values of N_A and N_D, respectively. With reverse bias the conduction band edge on the *n*-side (E_{CN}) drops below the valence-band edge on the *p*-side (E_{VP}), providing empty states in the conduction band of the *n*-material into which

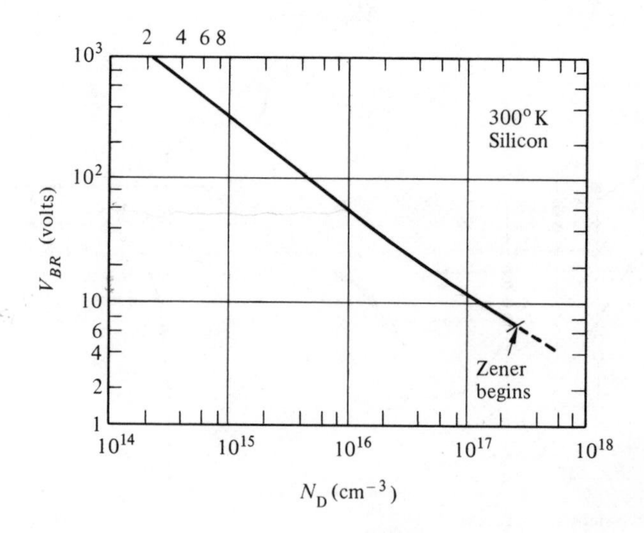

Fig. 4.3 Avalanche breakdown voltage for silicon p^+-n diode.

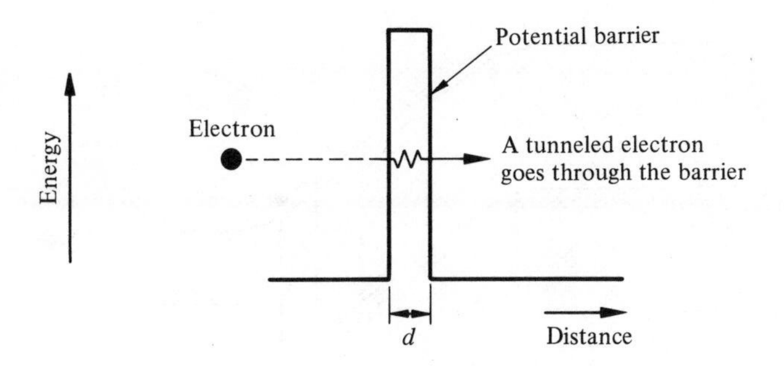

Fig. 4.4 Electron tunneling.

electrons may tunnel. Also, note that the filled electron states on the p-side provide the electrons for tunneling. Once tunneling begins, any additional reverse bias will increase the difference ($E_{VP} - E_{CN}$) and make more electrons available to tunnel into more empty states, thereby yielding even larger reverse currents.

4.1.3 Generation in *W*

The reverse current illustrated in Fig. 4.1 does not saturate at $-I_0$ as predicted by the ideal diode equation. In the development of the ideal diode equation it was assumed that the net generation–recombination rate in the depletion region was zero. When reverse biased, generation in the depletion region actually dominates, because the carrier concentrations are less than their thermal equilibrium values. Each electron–hole pair generated in W falls down the potential hill as illustrated in Fig. 4.6. Each thermally generated carrier

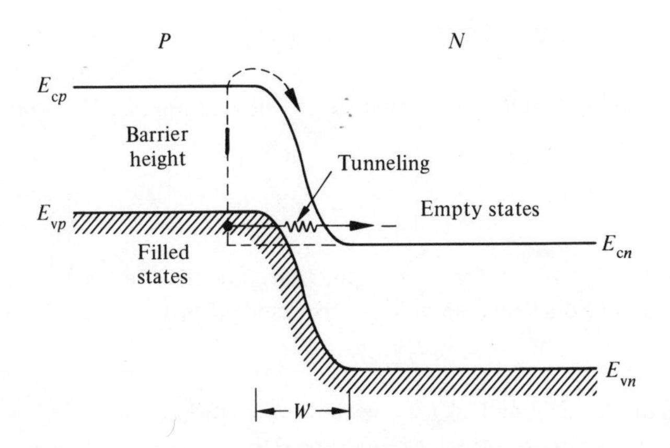

Fig. 4.5 Zener breakdown.

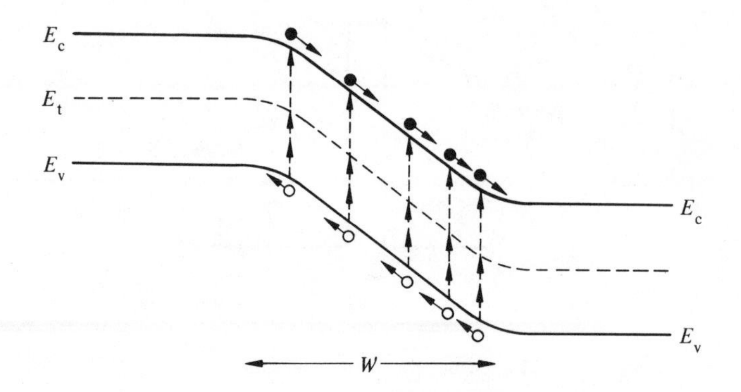

Fig. 4.6 Generation in the depletion region.

contributes to the current in the external circuit. Therefore a current in excess of $-I_0$ flows through the diode.

The generation rate of electron–hole pairs (G, #/cm^3-sec) can be used to calculate the additional reverse current. The number of additional carriers created per second is the volume times the generation rate. The additional current produced is

$$I_{\text{R-G}} = -qA \int_{-x_p}^{x_n} G\,dx \tag{4.8}$$

The generation rate in the depletion region is approximately a constant,

$$G = \frac{n_i}{2\tau_0} \tag{4.9}$$

where τ_0 is defined as an *effective lifetime* according to Eq. (4.10),

$$\tau_0 = \frac{\tau_n + \tau_p}{2} \tag{4.10}$$

The total generation current is then obtained by substituting Eq. (4.9) into Eq. (4.8) and integrating,

$$I_{\text{R-G}} = -qA\frac{n_i}{2\tau_0}W \tag{4.11a}$$

Remember that $W \propto |V_A|^{1/2}$ for a step junction and therefore the generation current increases with additional reverse voltage, as depicted in Fig. 4.1. The total current is

$$I = I_0(e^{qV_A/kT} - 1) + I_{\text{R-G}} \tag{4.11b}$$

The ideal diode equation leakage current ($-I_0$) is proportional to n_i^2, as evidenced by Eq. (3.53). The generation current is proportional to n_i and for silicon at room temperature

n_i is about $10^{10}/\text{cm}^3$. A p^+-n diode has typical values of $W \cong 10^{-4}$, $\tau_0 \cong \tau_p$, $L_P \cong 10^{-2}$, and $N_D = 10^{14}/\text{cm}^3$. Comparing the reverse current components,

$$\frac{n_i^{!}}{2\tau_0}W \cong \frac{10^{10} \times 10^{-4}}{2 \times \tau_0} = \frac{5 \times 10^5}{\tau_0}$$

and

$$\frac{n_i^2}{N_D}\frac{L_P}{\tau_p} \cong \frac{10^{20} \times 10^{-2}}{10^{14}\tau_p} \cong \frac{10^4}{\tau_p}$$

It is evident that for a silicon diode at room temperature the generation current is significantly larger. For a germanium diode, where n_i is about $10^{13}/\text{cm}^3$, the current terms become $(5 \times 10^8)/\tau_0$ for generation and $10^{10}/\tau_p$ for drift. A germanium diode, at room temperature under reverse bias, approximates the ideal diode equation much better than does a silicon device.

4.2 FORWARD BIAS DEVIATIONS FROM IDEAL

The ideal diode equation, derived in Chapter 3, is plotted in Fig. 4.7 for forward bias voltages on a semilogarithmic scale to emphasize the large range of current values. The parameter "n" is called the *ideality factor* and is a measure of how close to ideal were the conditions under which the physical device was fabricated. The ideality factor is incorporated into the ideal equation by replacing q/kT by q/nkT,

$$I = I_0(e^{qV_A/nkT} - 1) \tag{4.12}$$

Under any reasonable amount of forward bias, the exponential term is much greater than -1 and

$$I \cong I_0 e^{qV_A/nkT} \tag{4.13}$$

Taking a natural logarithm yields Eq. (4.14):

$$\ln I = \ln I_0 + \frac{qV_A}{nkT} \tag{4.14}$$

When Eq. (4.14) is plotted on a semilog scale with V_A as the abscissa, we get a result similar to the xy-relationship of Eq. (4.15), where "a" is the $x = 0$ intercept and "b" is the slope:

$$y = a + bx \tag{4.15}$$

Comparing Eq. (4.14) to Eq. (4.15), we see that the intercept is I_0, as illustrated in Fig. 4.7, and the slope is q/nkT. Therefore when $n = 1$ the physical data matches the ideal equation. For most present-day silicon devices, $1.0 \leq n \leq 1.06$ over 5 or 6 decades of current.

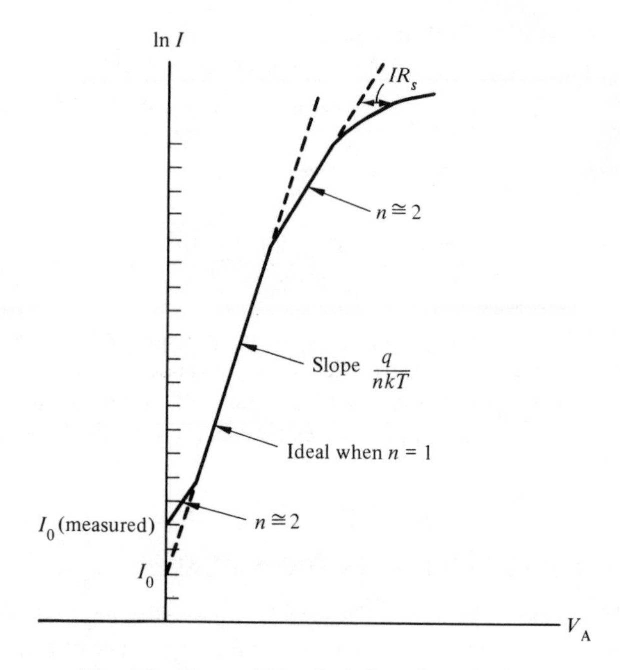

Fig. 4.7 Forward bias deviations from the ideal.

Three regions of nonideal behavior are illustrated in Fig. 4.7. At very small currents (near the extrapolated $V_A = 0$ intercept) the measured current is larger than the equation-predicted I_0. For large currents the slope decreases and eventually no specific slope can be determined.

4.2.1 Recombination in *W*

Consider the nonideal region at very small values of forward current. In this region of operation the injected carrier densities are relatively small and holes leaving the p-region travel through the depletion region on their way to be injected as minority carriers into the n-region. Simultaneously, electrons are traveling from the n-region through W to be injected into the p-region. The ideal diode equation assumed that no recombination–generation occurs in the depletion region. With the carrier numbers being greater than their thermal equilibrium values, recombination can occur in W. Each recombination event removes an excess electron and hole. Figure 4.8 is a representation of such an event, where the recombination takes place near the center of the depletion region. Remember that for the ideal diode, the currents injected were determined by the excess carriers at the edges of the depletion region. The excess carriers at the depletion region edges were determined by the applied voltage V_A. Therefore, at a particular value of applied voltage the carriers completely traversing W are those injected; in the present

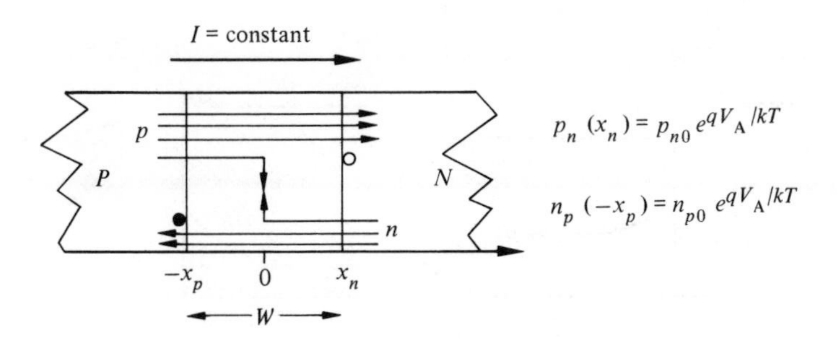

Fig. 4.8 Recombination in the depletion region.

case more carriers must enter W because some are lost to recombination. The result is a larger total current than predicted by the ideal diode equation at a fixed value of V_A.

The total current is constant throughout the diode; therefore it is also constant in the depletion region, even though part of the hole current entering W from the p-region never reaches the n-region, and similarly, some of the electrons entering W from the n-region never reach the p-region. The electrons recombine with the holes from the p-region as illustrated in Fig. 4.8, thereby forming a recombination current component of the total current. The recombination current component is added to the ideal diode diffusion currents.

At very low levels of current the recombination current component dominates in silicon diodes. It should be noted that at larger current levels the recombination current is still present, but is only a small fraction of the total current. Figure 4.7 illustrates these effects for a typical silicon device. A derivation of the recombination current is possible; however, we only present the result in terms of a modified diode equation, Eq. (4.16).

$$I = I_0(e^{qV_A/nkT} - 1) + q\frac{An_i}{2\tau_0}W(e^{qV_A/2kT} - 1) \tag{4.16}$$

When the recombination current dominates at low levels of current, the second term of Eq. (4.16) is larger than the first, giving the slope of $q/2kT$ as indicated in Fig. 4.7. Note that Eq. (4.16) is also valid for reverse bias.

4.2.2 High-Level Injection

The criterion for low-level injection is that the total minority carrier concentration always be much less than the equilibrium majority carrier concentration. For the *p-n* junction under forward bias, the largest minority concentration occurs at the depletion region edges; for example, see Fig. 4.9(a). At high-level injection the excess minority concentration approaches the majority concentration. If the bulk region is to maintain charge neutrality, the majority carrier concentration must also increase significantly above its equilibrium value as illustrated in Fig. 4.9(b). The original assumption of low-level injection resulted

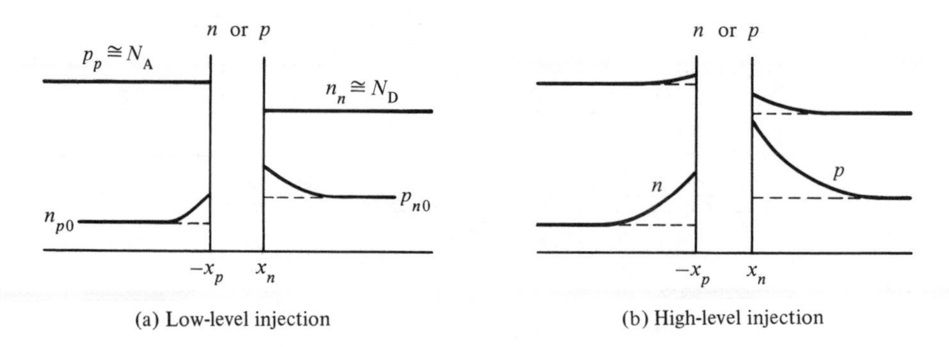

Fig. 4.9 Carrier concentrations: (a) low-level injection; (b) high-level injection.

in the definition of the recombination term to be approximated by $\Delta p/\tau_p$ and $\Delta n/\tau_n$. High-level injection requires a different recombination term with a redefinition of carrier lifetimes.

The net result of high-level injection is an increase in recombination that changes the ideality factor to about $n = 2$.

4.2.3 Bulk Region Effects

In the derivation of the ideal diode equation, it was assumed that the electric field in the bulk n and p regions was approximately zero and that no voltage drop existed across the ohmic contacts. For most modern devices these are good assumptions at the lower current levels. However, at large current levels the bulk resistance can produce a significant voltage drop and the applied voltage (V_A) is larger than the voltage across the depletion region. Also, the metal–silicon contacts can behave like small resistors, adding to the voltage drop across the n- and p-bulk regions. Usually these two effects are combined into a resistor R_S, the *series resistance*. Figure 4.7 illustrates its effect on the diode volt–ampere characteristic.

One method used to measure R_S is to plot the voltage drop from ideal (ΔV) versus current on a linear plot. If all the data points fall on a straight line, then R_S can be determined from the slope. The I-versus-V_A plots like Fig. 4.7 are routinely obtained by pulsing the current so as to avoid any self-heating effects on the results.

4.3 SUMMARY

The deviations in current and voltage of a physical diode from those of the ideal diode were considered first with reverse bias and then with forward bias. When reverse biased, the deviations were attributed to avalanche or Zener breakdown and to generation of electron–hole pairs in the depletion region. The forward biased deviations were due to recombination in the depletion region, high-level injection, and the ohmic resistances of the bulk regions and metal contacts.

Avalanche breakdown occurred at reverse bias voltages sufficiently large that the electric field in the depletion region exceeded its critical value. At the critical value of electric field, the carriers gain enough energy so that in colliding with a crystal atom they generate an electron–hole pair. The newly generated carriers also gain energy and cause additional avalanche generation until large currents are created.

Zener breakdown is obtained from heavily doped p^+-n^+ junctions, where electrons can tunnel directly from the valence band to the conduction band. A small change in voltage results in larger numbers of carriers being able to tunnel.

Thermal generation of electrons and holes in the depletion width of the reverse biased diode gives rise to a current component in addition to $-I_0$. The amount of generation controls the size of the current, that is, I_{R-G} is proportional to n_i and W.

Recombination of electrons with holes as they traverse the depletion region in a forward biased diode forces a larger current component than that predicted by the ideal diode equation. Only at small values of total current is the recombination in the W component of significance.

When the current density is quite large, either high-level injection and/or the series resistance of the device becomes important. High-level injection has the effect of changing the ideality factor to about two. The series resistance increases the voltage drop across the diode for a given current.

PROBLEMS

4.1 A p^+-n silicon step junction is doped $N_D = 10^{16}/\text{cm}^3$. Determine the approximate values of

(a) V_{BR};

(b) W at V_{BR};

(c) $\mathscr{E}_{max} = \mathscr{E}_{CR}$;

(d) If $N_D = 10^{17}/\text{cm}^3$, calculate parts (a), (b), and (c).

4.2 Sketch the electric field in the depletion region of an avalanching step junction showing $\mathscr{E}_{CR}$ and indicate where avalanching is occurring. Repeat for a larger reverse voltage.

4.3 A Si step junction with area 10^{-4}cm^2 is doped $N_A = 5 \times 10^{15}/\text{cm}^3$ and $N_D = 10^{15}/\text{cm}^3$; see Problem 3.1. If $\tau_n = 0.4$ μ sec and $\tau_n = 0.1$ μ sec, calculate:

(a) generation rate in W;

(b) generation current when $V_A = -0.1$ volts and $V_A = -10$ volts;

(c) I_0/I_{R-G} at $V_A = -10$.

4.4 The nonideal generation current in the forward biased diode is often modeled by two devices in parallel, as illustrated by Fig. P4.4. Plot on semilog (paper) the composite $I-V_A$ from 10^{-14} to 10^{-3} A, if $I_{01} = 10^{-15}$ A, $n_1 = 1$ and $I_{02} = 10^{-13}$ A, $n_2 = 2$.

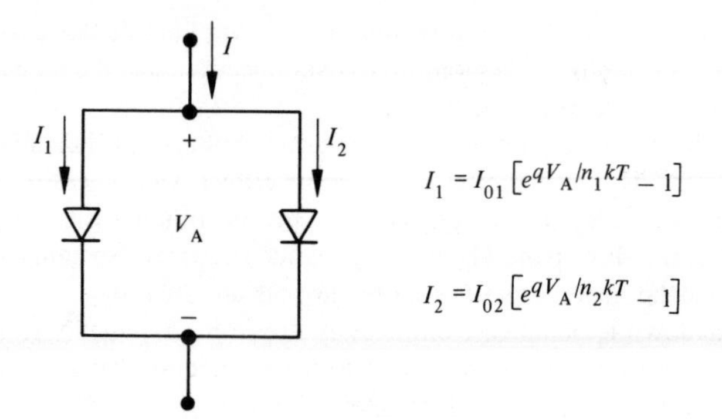

Fig. P4.4

4.5 For the diode of Problem 4.3 plot Eq. (4.16) on a semilog plot (assuming $n = 1.0$) from 250 mV to 600 mV; plot the diffusion current component, recombination component, and the total current. Is the recombination component a straight line? Explain.

4.6 If $R_S = 20\Omega$, at what current will the applied voltage deviate from the ideal by

(a) 10%. Let $I_0 = 10^{-14}$ A.

(b) If $R_S = 2\Omega$ repeat part (a).

4.7 Explain where, in a p^+-n diode, high-level injection will first occur and how it affects the $\ln I$-versus-V_A plot. What is the relationship between N_D of the device and the current at the onset of high-level injection?

5 / *p-n* Junction Admittance

The previous two chapters have described the response of the *p-n* junction to a dc voltage. The present chapter will investigate the response of the diode to a small-signal (sine wave) voltage superimposed upon the dc voltage. Usually the signal response is described in terms of the *small-signal admittance* (Y), which has a real part called the *conductance* and an imaginary part called the *susceptance*. The term "small-signal" implies that the peak values of the signal current and voltage are much smaller than the dc values. Typically this means a signal voltage of several millivolts or less.

5.1 REVERSE BIASED JUNCTION ADMITTANCE

The small-signal admittance of the reverse biased *p-n* junction is modeled by a conductance and capacitive susceptance in parallel as illustrated in Fig. 5.1. When a small-signal, sine wave voltage is applied to the diode in addition to the reverse biased dc voltage, as illustrated in Fig. 5.2, the admittance can be expressed as Eq. (5.1),

$$Y = G + j\omega C \tag{5.1}$$

For frequencies sufficiently low, where the carrier response time is much shorter than the period of the signal, Eq. (5.1) simplifies to Eq. (5.2):

$$Y = G_0 + j\omega C_J \tag{5.2}$$

Here G_0 is the low-frequency conductance which is independent of ω, but as will be derived, is dependent on the dc operating variables V_A and I. The *depletion layer* or, as often termed, *junction capacitance* C_J, is also dependent upon the dc variables and is independent of frequency. The remainder of this section is devoted to obtaining a physical insight into how the junction responds to the signal and to obtaining expressions for G_0 and C_J.

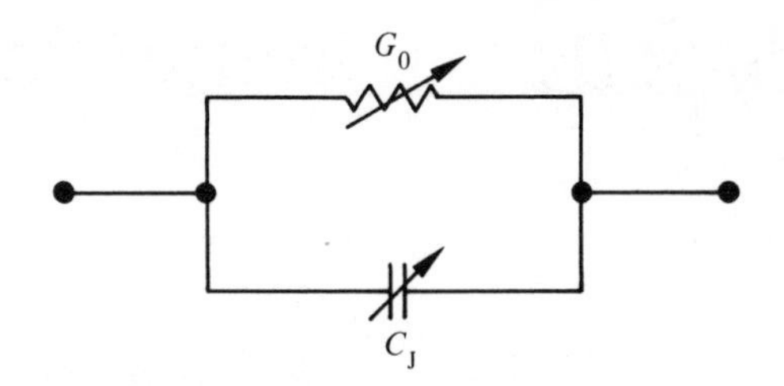

Fig. 5.1 Small signal equivalent circuit for the reverse biased diode.

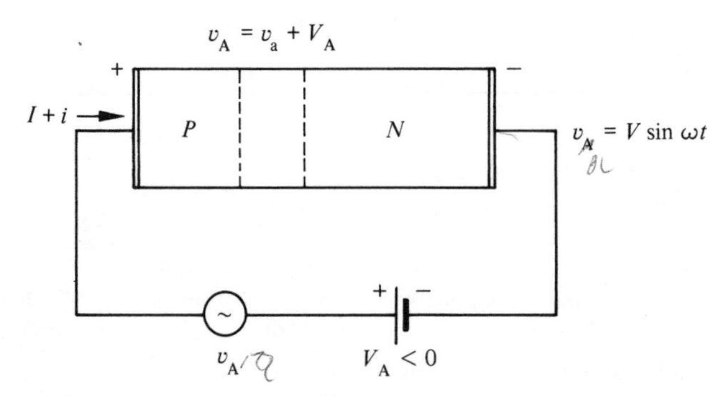

Fig. 5.2 Small signal ac voltage plus the dc bias voltage.

5.1.1 Junction Capacitance

Unlike parallel plate capacitance, which is constant, the junction capacitance changes with the applied dc voltage as illustrated in Fig. 5.3. The capacitance decreases as V_A becomes more negative.

To explain this phenomenon, consider the depletion region at some fixed value of reverse voltage. For a step junction the depletion width (W) is determined by Eq. (2.51), repeated here for convenience.

$$W = \left[\frac{2K_S\varepsilon_0}{q}(V_{bi} - V_A)\frac{(N_A + N_D)}{N_A N_D}\right]^{1/2} \tag{5.3}$$

With the signal superimposed, V_A is replaced by $(V_A + v_a)$ in Eq. (5.3) and W increases or decreases by an increment ΔW. For small signals, $|v_a| \ll |V_A|$, and therefore $\Delta W \ll W$. Nevertheless, charge is added to or subtracted from the depletion region edges in response to v_a. When $v_a > 0$, W decreases by adding holes to the p-region and electrons to the n-region as illustrated in Fig. 5.4(b). As W becomes smaller, it must do so by covering the $-qN_A$ ion with holes—the majority carrier. Similarly the $+qN_D$ charge is covered, or neutralized, with majority carrier electrons in the n-region. The net charge increments for $v_a > 0$ are illustrated in Fig. 5.4(c).

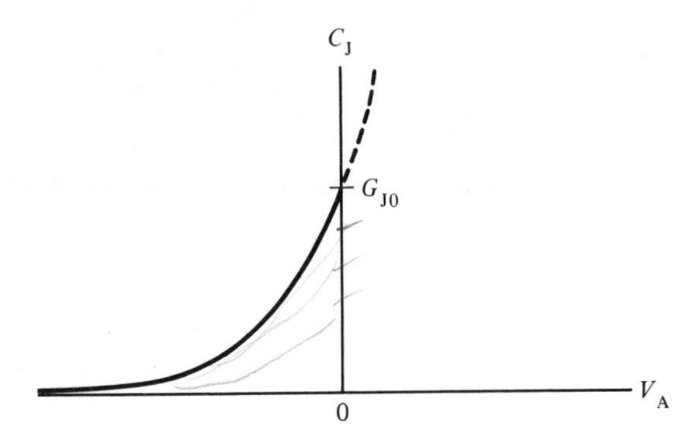

Fig. 5.3 Junction or depletion capacitance.

When $v_a < 0$, W must become incrementally larger by depleting holes from the p-region and electrons from the n-region as illustrated in Fig. 5.4(a) and (b). The charge increments for this case are illustrated in Fig. 5.4(d). Two important observations must be emphasized. First, the charge being moved is always a majority carrier charge. Majority carrier charges respond to a voltage change at roughly the dielectric relaxation time of the material. In silicon, at normal doping levels, the majority carrier response time is from about 10^{-10} to 10^{-12} sec. With such short response times, the phenomena will be independent of the frequency of v_a up to very high frequencies. The second observation is that the incremental charge diagrams of Fig. 5.4(c) and (d) are similar to the charge fluctuations on the parallel plates of a capacitor with area (A) and separation (W). Because of this similarity between the reverse biased diode and the capacitor, the depletion (C_J) is obtained from the parallel plate capacitance formula as Eq. (5.4). Equation (5.4) is valid provided W is essentially fixed; that is, ΔW must be very small compared to W:

$$C_J = \frac{K_S \varepsilon_0 A}{W} \tag{5.4}$$

For small signals $|v_a| \ll |V_A|$, and when $v_A = V_A + v_a$ is substituted into Eq. 5.3 for V_A, then ΔW is $\ll W$. Substituting Eq. (5.3) into Eq. (5.4), remembering that $V_A < 0$, we obtain Eq. (5.5),

$$C_J = \frac{K_S \varepsilon_0 A}{\left[\dfrac{2K_S \varepsilon_0}{q}(V_{bi} - V_A)\dfrac{(N_A + N_D)}{N_A N_D}\right]^{1/2}} \tag{5.5}$$

An examination of Eq. (5.5) yields the following facts.

1. As indicated in Fig. 5.3, C_J decreases as V_A becomes more negative because W

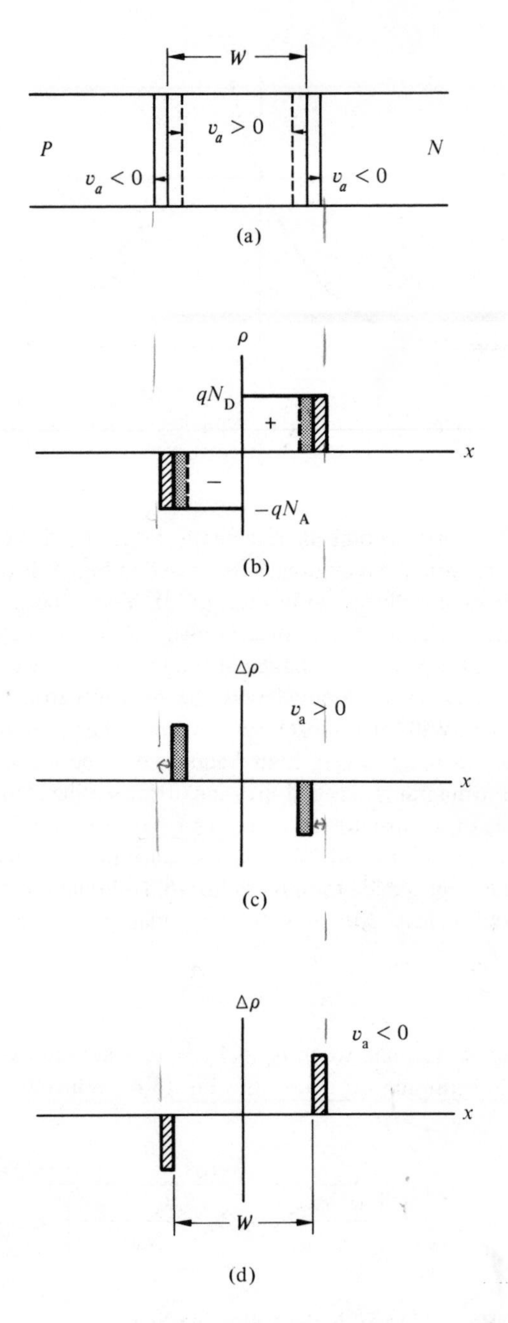

Fig. 5.4 Charge increments for C_j the depletion capacitance.

increases as discussed in Chapter 2, that is,

$$C_J \propto \frac{1}{|V_A|^{1/2}}, \qquad \text{if } |V_A| \gg V_{bi}$$

2. If N_A or N_D is increased, W decreases, thereby increasing C_J.

3. For a p^+-n junction $C_J \propto N_D^{1/2}$. For a p-n^+ junction $C_J \propto N_A^{1/2}$.

4. If the junction were a linear-graded junction rather than an abrupt junction, then at $|V_A| \gg V_{bi}$, W would vary as the one-third power of V_A rather than the one-half power; that is,

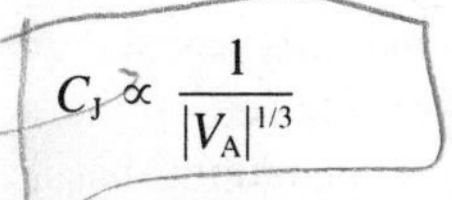

$$C_J \propto \frac{1}{|V_A|^{1/3}}$$

(See Chapter 2, Eq. (2.57) for further details.)

Equation (5.5) is often generalized and written in terms of the *zero-biased* junction capacitance (C_{J0}). In Eq. (5.5) let $V_A = 0$; then the definition of C_{J0} is as follows:

$$C_{J0} = C_J|_{V_A=0} = \frac{K_S\varepsilon_0 A}{\left[\frac{2K_S\varepsilon_0}{q}V_{bi}\frac{(N_A + N_D)}{N_A N_D}\right]^{1/2}} \tag{5.6}$$

Factoring out C_{J0} from Eq. (5.5) yields Eq. (5.7):

$$C_J = \frac{K_S\varepsilon_0 A}{\left[\frac{2K_S\varepsilon_0}{q}V_{bi}\frac{(N_A + N_D)}{N_A N_D}\right]^{1/2}\left(1 - \frac{V_A}{V_{bi}}\right)^{1/2}} = \frac{C_{J0}}{\left(1 - \frac{V_A}{V_{bi}}\right)^{1/2}} \tag{5.7}$$

Finally, replacing the one-half power in Eq. (5.7) by "m" yields Eq. (5.8). The primary reason for introducing the generalized "m" is for handling experimental data on p-n junctions where the abruptness is usually unknown.

$$C_J = \frac{C_{J0}}{\left[1 - \frac{V_A}{V_{bi}}\right]^m}, \qquad 1/3 \le m \le 1/2 \tag{5.8}$$

For a linearly graded junction $m = 1/3$; for a step junction $m = 1/2$. Most real junctions, fabricated by standard procedures, are found to have an "m" of between 1/3 and 1/2.

It should be noted that Eq. (5.8) is applicable to small forward biased voltages of less than V_{bi}. Figure 5.3 illustrates that when the diode is forward biased, C_J increases very rapidly.

The applications of junction capacitors are extensive. In bipolar integrated circuits the reverse biased p-n junction is used to isolate the transistors and resistors from each other. Almost every FM tuner and TV set uses the dc voltage-variable capacitor in automatic tuning circuits; it has no moving parts and the capacitance changes in response to a dc level.

5.1.2 Conductance

The small-signal conductance of the reverse biased junction is derived under the assumption that the carriers are able to respond to the signal quasi-statically; that is, the carriers return to near steady state in much less time than the period of the signal. In the case of reverse bias this means that signal frequencies are less than about 100 MHz for most devices, since the carriers responding are the majority carriers.

Under the assumption of majority carrier quasi-static response, the diode reacts instantaneously to a signal superimposed on the dc operating variables. The diode can be represented by the ideal diode equation; that is, if the diode current is a function of the dc voltage V_A that is perturbed by a small signal v_a, Eq. (5.9) is modified as per Eq. (5.10).

$$I(V_A) = I_0[e^{qV_A/kT} - 1] \tag{5.9}$$

$$I(V_A + v_a) = I_0[e^{q(V_A + v_a)/kT} - 1] \tag{5.10}$$

and the signal current, i, is defined as Eq. (5.11)

$$i = I(V_A + v_a) - I(V_A) \tag{5.11}$$

Equation (5.10) can be expanded into the Taylor series of Eq. (5.12) and, because $v_a \ll V_A$, only the first two terms need be retained, giving Eq. (5.13),

$$f(x + h) = f(h) + x\frac{df}{dh} + \cdots \tag{5.12}$$

$$I(v_a + V_A) \cong I(V_A) + v_a\frac{dI}{dV_A} \tag{5.13}$$

Therefore the signal current is obtained from Eq. (5.11) as Eq. (5.14),

$$i = v_a\frac{dI}{dV_A} = \left[\frac{dI}{dV_A}\right]v_a \tag{5.14}$$

The low-frequency ($\omega \to 0$) conductance is defined from Eq. (5.14) as

$$G_0 = \frac{i}{v_a} = \frac{dI}{dV_A} \quad \text{mhos} \tag{5.15}$$

By differentiating the ideal diode equation, Eq. (5.9), we obtain the low-frequency conductance as Eq. (5.16),

$$G_0 = \frac{dI}{dV_A} = I_0\frac{q}{kT}e^{qV_A/kT} = \frac{q}{kT}(I + I_0) \tag{5.16}$$

The last form of Eq. (5.16) was obtained by adding I_0 to both sides of Eq. (5.9). A common model element is the *dynamic resistance*, defined as the reciprocal of G_0,

$$r = \frac{1}{G_0} = \frac{kT}{q(I + I_0)} \quad \text{ohm} \tag{5.17}$$

Note that for several tenths of a volt of reverse bias, $I \rightarrow -I_0$ and $r \rightarrow \infty$.

For the reverse bias case of a silicon diode at room temperature, as discussed in Chapter 4, the generation current in W dominates conduction. Applying Eq. (5.15) to that case yields

$$G_0 = \frac{q}{kT}(I + I_0) - \frac{qA}{2\tau_0} n_i \frac{dW}{dV_A} \tag{5.18}$$

Differentiating Eq. (2.51) results in Eq. (5.19)

$$\frac{dW}{dV_A} = \left[\frac{2K_S\varepsilon_0}{q}\frac{(N_A + N_D)}{N_A N_D}\right]^{1/2} \frac{(1/2)(-1)}{(V_{bi} - V_A)^{1/2}} \tag{5.19}$$

The low-frequency reverse bias conductance is then given by Eq. (5.20)

$$G_0 = \frac{q}{dT}(I + I_0) + \frac{qAn_i}{4\tau_0}\left[\frac{2K_S\varepsilon_0}{q}\frac{(N_A + N_D)}{N_A N_D}\frac{1}{(V_{bi} - V_A)}\right]^{1/2} \tag{5.20}$$

Note that the nonideal silicon diode has a finite conductance that depends on the generation current originating in the depletion region. Equation (5.16) and Eq. (5.20) for G_0 are also interpreted as the slope of the I–V_A curve. Any phenomenon that increases the slope increases the conductance. One final observation is that Eq. (5.16) will be shown in the next section to be valid for forward bias. The validity is, however, at frequencies considerably lower than for the reverse bias case.

5.2 FORWARD BIASED JUNCTION ADMITTANCE

The junction diode under forward bias conditions ($V_A > 0$) and perturbed by a small signal sinusoid is also modeled by an admittance. In addition to the depletion region capacitance C_J (which is a result of majority carrier response), the minority carrier response yields a *diffusion capacitance* C_D. The minority carriers also contribute to the conductance (G) of the junction admittance. Figure 5.5 illustrates the signal equivalent circuit where R_s is the series resistance of the diode due to the bulk regions and/or the ohmic contacts. In general, the conductance and diffusion capacitance are a function of the signal frequency and the dc operating point variables, as will become evident by the following derivation. Let's consider the response of the minority carriers.

The minority carrier concentrations for a p-n junction forward biased with a dc voltage $V_A > 0$ and perturbed by a small signal sinusoid are illustrated in Fig. 5.6. The average minority carrier concentration is a result of the voltage V_A across the depletion region. Increments above and below the average are due to the signal v_a applied to the depletion region edges. As v_a becomes positive and negative the carrier distributions at x_n and $-x_p$

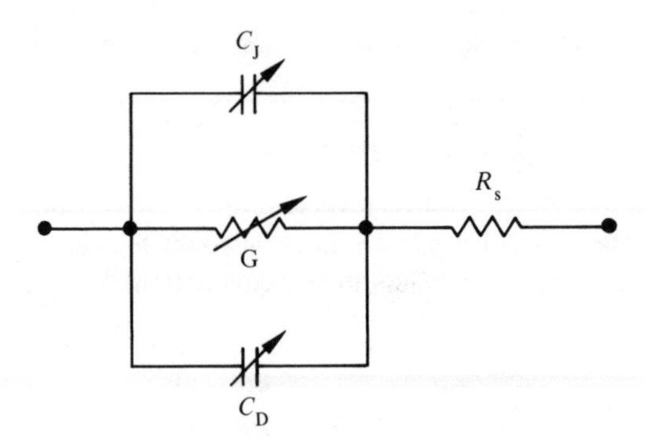

Fig. 5.5 Small signal model of the diode.

change. Because diffusion is a relatively slow process compared with the signal frequency, the charge distribution increment propagates into the bulk region, fluctuating above and below the average (dc) carrier distribution. Figure 5.6 illustrates the fluctuation of the electrons and holes for one instant of time.

A dynamic picture of Fig. 5.6 can be conceptualized by visualizing that you are holding one end of a long rope. When held fixed in position it drops from your hand to the floor in a manner similar to the dc steady-state, minority carrier distribution. If the held end of the rope is oscillated rapidly, the rope propagates the pulsation down its length until the pulse eventually decays away on the floor.

The diode minority carrier distribution propagates a signal current in response to the signal voltage. We should therefore expect an associated junction admittance that has a conductance and a susceptance, both of which will be functions of the dc operating point

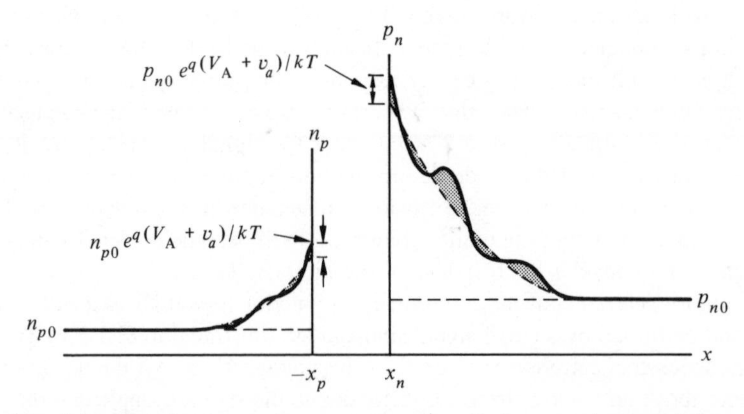

Fig. 5.6 High-frequency charge storage p^+-n diode at one instant of time.

and the signal frequency. Figure 5.6 illustrates that the minority carrier distributions are a function of time and space; that is $p_n(x, t)$.

To derive the diffusion admittance of a junction (the minority carrier response), let us assume a p^+-n junction and therefore concern ourselves only with $p_n(x, t)$. A similar derivation holds for $n_p(x, t)$. The minority carrier continuity equation for a uniformly doped n-region is the starting point, repeated here for the reader's convenience from Vol. I, Chapter 3,

$$\frac{\partial \Delta p_n(x,t)}{\partial t} = D_P \frac{\partial^2 \Delta p_n(x,t)}{\partial x^2} - \frac{\Delta p_n(x,t)}{\tau_p} \tag{5.21}$$

Figure 5.6 indicates that the signal oscillates the hole distribution around its average (dc) value. This allows the solution to be broken up into a solution for the dc component plus the signal components, as indicated by Eq. (5.22):

$$\Delta p_n(x,t) = \overline{\Delta p_n}(x) + \tilde{p}_N(x,t) \tag{5.22}$$

Substituting Eq. (5.22) into Eq. (5.21) yields Eq. (5.23),

$$\frac{\partial \overline{\Delta p_n}(x)}{\partial t} + \frac{\partial \tilde{p}_N(x,t)}{\partial t} = D_P \frac{\partial^2 \overline{\Delta p_n}(x)}{\partial x^2} + D_P \frac{\partial^2 \tilde{p}_N(x,t)}{\partial x^2} - \frac{\overline{\Delta p_n}(x)}{\tau_p} - \frac{\tilde{p}_N(x,t)}{\tau_p} \tag{5.23}$$

Since $\overline{\Delta p_n}(x)$ is not a function of time, the first term of Eq. (5.23) is equal to zero. Separately equating the coefficients of the average terms and the signal terms results in Eqs. (5.24) and (5.25).

$$0 = D_P \frac{\partial^2 \overline{\Delta p_n}(x)}{\partial x^2} - \frac{\overline{\Delta p_n}(x)}{\tau_p} \tag{5.24}$$

$$\frac{\partial \tilde{p}_N(x,t)}{\partial t} = D_P \frac{\partial^2 \tilde{p}_N(x,t)}{\partial x^2} - \frac{\tilde{p}_N(x,t)}{\tau_p} \tag{5.25}$$

Equation (5.24) was solved in Chapter 3 and results in the dc solution, in that chapter, led to the I versus V_A relationship known as the ideal diode equation. Let us turn our attention to the solution of Eq. (5.25), which will lead us to the small-signal admittance, our ultimate goal.

An examination of Eq. (5.25) shows it to be a type of differential equation that can be solved as a product solution (similar to the wave equation in electromagnetic field theory). The solution is broken into the signal part that is only a function of x and the signal part that is only a function of time.

$$\tilde{p}_N(x,t) = \hat{p}_N(x) f(t) \tag{5.26}$$

If we assume the forcing function to be a sine or cosine function, then $f(t) = e^{j\omega t}$, similar to the phasor concept from circuit theory,

$$\tilde{p}_N(x,t) = \hat{p}_N(x) e^{j\omega t} \tag{5.27}$$

Therefore, if Eq. (5.27) is an assumed solution it must satisfy Eq. (5.25); that is, performing the differentiations of Eq. (5.25), using Eq. (5.27), must yield

$$\frac{\partial \tilde{p}_N(x,t)}{\partial t} = j\omega \hat{p}_N(x) e^{j\omega t} \tag{5.28}$$

$$\frac{\partial^2 \tilde{p}_N(x,t)}{\partial x^2} = \frac{d^2 \hat{p}_N(x)}{dx^2} e^{j\omega t} \tag{5.29}$$

Substitution into Eq. (5.25) gives Eq. (5.30), and cancelling the $e^{j\omega t}$ terms before collecting the $\hat{p}_n(x)$ terms yields Eq. (5.31),

$$j\omega \hat{p}_N(x) e^{j\omega t} = D_P \frac{d^2 \hat{p}_N(x)}{dx^2} e^{j\omega t} - \frac{\hat{p}_N(x) e^{j\omega t}}{\tau_p} \tag{5.30}$$

or

$$j\omega \hat{p}_N(x) + \frac{\hat{p}_N(x)}{\tau_p} = D_P \frac{d^2 \hat{p}_N(x)}{dx^2} = \hat{p}_N(x) \left[\frac{1}{\tau_p} + j\omega \right] \tag{5.31}$$

Dividing by D_P results in an equation very similar to the differential equation for the dc solution of the minority carrier concentrations,

$$\frac{d^2 \hat{p}_N(x)}{dx^2} = \hat{p}_N(x) \left[\frac{1 + j\omega\tau_p}{D_P \tau_p} \right] \tag{5.32}$$

To make Eq. (5.32) have the same form as the dc case, we define the complex diffusion length for holes as Eq. (5.33),

$$[L_P^*]^2 = \frac{D_P \tau_p}{1 + j\omega\tau_p} = \frac{L_P^2}{1 + j\omega\tau_p} \tag{5.33}$$

Then Eq. (5.32) can be written as Eq. (5.34)

$$\frac{d^2 \hat{p}_N(x)}{dx^2} = \frac{\hat{p}_N(x)}{[L_P^*]^2} \tag{5.34}$$

The solution of Eq. (5.34), by analogy with the dc solution of Eq. (5.24) [or Eq. (3.36)], is

$$\hat{p}_N(x) = B_1 e^{x/L_P^*} + B_2 e^{-x/L_P^*} \tag{5.35}$$

As a boundary condition we know that $|\hat{p}_N(x)|$ cannot grow without bound; that is, as $x \to \infty$ $|\hat{p}_N(x)| \to 0$, and therefore B_1 must be zero. At the depletion region edge, the signal component of the hole distribution is controlled by the applied signal voltage. To evaluate the constant B_2 we need to know the value of $\hat{p}_N(0)$. The total carrier distribution is determined by the total voltage applied, as given by Eq. (5.36), which is the same as Eq. (3.30), with $V_A \to V_A + v_a(t)$ and the x' axis starting at x_n,

$$p_n(0,t) = p_{n0} \exp \left[\frac{q(V_A + v_a(t))}{kT} \right] \tag{5.36}$$

The signal is small; that is $v_a(t) \ll V_A$, and the series expansion of the exponential can be applied to Eq. (5.36) as

$$e^x \cong 1 + x + \cdots, \qquad \text{for } x \ll 1 \tag{5.37}$$

or

$$p_n(0, t) \cong p_{n0}e^{qV_A/kT}\left[1 + \frac{qv_a(t)}{kT}\right]$$

$$= p_{n0}e^{qV_A/kT} + p_{n0}e^{qV_A/kT}\frac{qv_a(t)}{kT} \tag{5.38}$$

Note that the signal part is the last term of Eq. (5.38), and therefore

$$\hat{p}_n(0) = p_{n0}e^{qV_A/kT}\frac{qv_a}{kT} = B_2 \tag{5.39}$$

where the constant B_2 was evaluated using Eq. (5.35).

The signal current is obtained from the diffusion current formula of Eq. (5.40),

$$i = -qAD_P\frac{d\hat{p}_N(x)}{dx}\bigg|_{x=0} = \frac{qAD_P}{L_P^*}p_{n0}\, e^{qV_A/kT}\frac{qv_a}{kT} \tag{5.40}$$

The junction admittance is then

$$Y = A\frac{q}{kT}\left[q\frac{D_P}{L_P^*}p_{n0}\right]e^{qV_A/kT} \tag{5.41}$$

From Eq. (5.33)

$$L_P^* = \frac{L_P}{\sqrt{1 + j\omega\tau_p}} \tag{5.42}$$

and Eq. (5.41) can be rewritten as Eq. (5.43)

$$Y = \frac{qA}{kT}\left[q\frac{D_P p_{n0}}{L_P}\sqrt{1 + j\omega\tau_p}\right]e^{qV_A/kT} \tag{5.43}$$

Note that the square root of a complex number is another complex number, and Eq. (5.43) will have a real part which is the conductance G and an imaginary part which will be the capacitive susceptance ωC_D. The *diffusion capacitance* is then defined as

$$C_D = \frac{\text{Imaginary part of } Y}{\omega} \tag{5.44}$$

Equation (5.43) can easily be extended to include the minority carrier electron contribution by using the idea of "complements" introduced earlier. The full derivation for electrons is identical to that for the holes. Therefore the total diffusion admittance of a p-n junction is

$$Y = \frac{qA}{kT}\left[q\frac{D_P}{L_P}p_{n0}\sqrt{1 + j\omega\tau_p} + q\frac{D_N}{L_N}n_{p0}\sqrt{1 + j\omega\tau_n}\right]e^{qV_A/kT} \tag{5.45}$$

where the real part of Eq. (5.45) is the junction conductance due to holes in the n-region and electrons in the p-region. Similarly, the diffusion capacitance, due to holes and electrons, is obtained by applying Eq. (5.44) to Eq. (5.45).

5.3 LIMITING CASES

5.3.1 p^+-n Diode Admittance

To develop a number of important concepts and practical results, let us return to a simpler case, that of the p^+-n device. The doping asymmetry of the heavily doped region makes $n_{p0} \ll p_{n0}$. If $\omega \to 0$, Eq. (5.43) becomes the definition of the low-frequency conductance, G_0

$$Y\big|_{\omega = 0} = \frac{q}{kT}\left[qA\frac{D_P p_{n0}}{L_P}e^{qV_A/kT}\right] = G_0 \tag{5.46}$$

Now Eq. (5.43) can be simplified to Eq. (5.47),

$$Y = G_0\sqrt{1 + j\omega\tau_p} \qquad \text{mhos} \tag{5.47}$$

To evaluate Eq. (5.47), it is necessary to take the square root of a complex number. This means converting $1 + j\omega\tau_p$ to polar coordinates, taking the square root of the magnitude and one-half the angle, followed by a conversion back to rectangular coordinates, yielding a real part and an imaginary part.

The frequency dependence of G and C_0 is most easily observed by plotting Eq. (5.47) for specific values of $\omega\tau_p$. Figure 5.7 illustrates that the conductance begins to increase with frequency, above the low-frequency value, when $\omega\tau$ is about 0.5, while the diffusion capacitance decreases at the higher frequencies.

The dependence of the conductance and diffusion capacitance on the dc operating point variable is most easily obtained from the ideal diode equation and Eq. (5.47). For any reasonable degree of forward bias

$$G \propto e^{qV_A/kT} \propto I \tag{5.48}$$

and

$$C_D \propto G_0 \propto I \tag{5.49}$$

The low-frequency conductance G_0 of Eq. (5.46) is identical to the case of reverse bias of Eq. (5.15). Note that the bracketed term of Eq. (5.46) contains I_0 from the ideal diode equation for a p^+-n device,

$$\frac{qAD_P p_{n0}}{L_P}e^{qV_A/kT} = I_0 e^{qV_A/kT} = (I + I_0) \tag{5.50}$$

therefore

$$G_0 = \frac{q}{kT}(I + I_0) = \frac{dI}{dV_A} \tag{5.51}$$

From Fig. 5.7, this means that Eq. (5.51) is applicable for $\omega\tau_p < 0.5$.

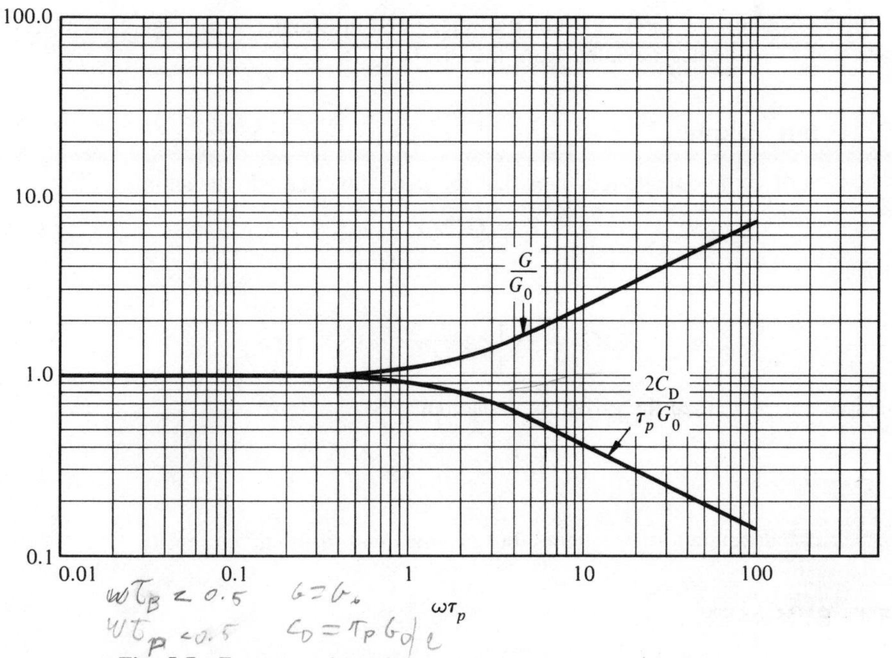

Fig. 5.7 Frequency dependence of G and C_D for a p^+-n junction.

5.3.2 p^+-n Diode, $\omega\tau_p \ll 1$

Equation (5.47) can be expanded in a Taylor series of Eq. (5.12), the result of which is Eq. (5.52)

$$Y \cong G_0\left[1 + \frac{j\omega\tau_p}{2}\right] = G_0 + j\omega\frac{G_0\tau_p}{2}, \qquad \text{for } \omega\tau_p \ll 1 \tag{5.52}$$

and the diffusion capacitance at low frequencies becomes

$$C_D \cong \frac{G_0\tau_p}{2} \tag{5.53}$$

which is frequency independent as illustrated by Fig. 5.7. Again note the dependence of C_D on $G_0 \propto 1$. Also the larger τ_p, the more minority carrier charge-storage phenomena; that is, a larger C_D.

5.3.3 Series Resistance

Most practical devices have some small but finite resistance in the bulk regions. This resistance is in series with the device. The ohmic contacts, the metal to silicon, also add resistance. Figure 5.5 illustrates the sum of these two as R_S. It should be noted that R_S

is fundamentally the same resistance as discussed in Chapter 4, at large current levels, where the real device deviated from the ideal.

5.3.4 p-n$^+$ Diode

The idea of complements indicates that the p-n^+ junction admittance is

$$Y = G_0\sqrt{1 + j\omega\tau_n} \tag{5.54}$$

where

$$G_0 = \frac{q}{kT}\left[qA\frac{D_N n_{p0}}{L_N}\right]e^{qV_A/kT} \propto I \tag{5.55}$$

and as $\omega\tau_n \ll 1$, then the diffusion capacitance is

$$C_D \cong \frac{G_0\tau_n}{2}, \quad \text{for } \omega\tau_n \ll 1 \tag{5.56}$$

which is independent of frequency but proportional to the dc current I.

5.4 SUMMARY

The junction admittance for the reverse biased diode was developed by considering the superimposed, small signal (v_a) on the dc voltage and its effect on the current. At frequencies where the majority carriers have ample time to respond to the signal voltage, the junction was modeled by a *junction capacitance* and a conductance. The junction capacitance decreased with larger reverse biases, as did the conductance. An increase in doping densities, N_A and/or N_D, increased the junction capacitance by making the depletion width smaller.

The case of the diode in forward bias resulted in a diffusion admittance that is dependent on the signal frequency. At very low frequencies the conductance and diffusion capacitance are constant. For frequencies such that $\omega\tau$ approaches 0.5, the conductance increases, and the diffusion capacitance decreases with increasing frequency. The admittance increases exponentially with V_A and is proportional to the dc current I. Both G and C_D are therefore proportional to the dc operating point current.

The total forward biased, small-signal equivalent circuit contains C_J, G, and C_D in parallel, with R_s in series with the combination.

PROBLEMS

5.1 A p-n silicon step junction doped $N_A = 10^{17}/\text{cm}^3$ and $N_D = 5 \times 10^{15}/\text{cm}^3$ has $\tau_n = 10^{-8}$ sec and $\tau_p = 10^{-7}$ sec, with $A = 10^{-4}\text{cm}^2$, $V_{bi} = 0.7603$, $\mu_n = 801\ \text{cm}^2/\text{V-sec}$, and $\mu_p = 438\ \text{cm}^2/\text{V-sec}$ [see Problem 3.2]. Calculate

(a) The junction capacitance at

1. $V_A = 0$;
2. $V_A = -0.3801$ volts $(-V_{bi}/2)$;
3. $V_A = -10$ volts.

(b) The diffusion capacitance ($\omega\tau < 0.1$ for n and p) $V_{bi} = 0.7603$

1. $V_A = V_{bi}/2$;
2. $V_A = .9V_{bi}$;
3. junction conductance at V_A of (1) and (2).

(c) Discuss which carrier type dominates the diffusion capacitance and the junction capacitance. Is $C_D \propto I$?

5.2 If in Problem 5.1, $\omega = 10^7$ rad/sec, calculate the diffusion capacitance and the conductance at $V_A = 0.9V_{bi}$. Compare the results with Problem 5.1, part (b) 2.

5.3 The junction (depletion capacitance) for an abrupt junction has the experimental values of

C_j (pF)	V_A
3.993	−.5
3.420	− 1
2.764	− 2
2.381	− 3
2.123	− 4
1.934	− 5

(a) Carefully plot $1/C^2$ versus V_A and determine C_{J0} and V_{bi} from the graph.

(b) Outline the procedure for C_{J0} and V_{bi} for the linear graded junction.

5.4 For a p^+-n junction sketch:

(a) C_{J0} versus N_D;

(b) C_D versus τ_p when $\omega\tau_p < 0.1$;

(c) $\omega\tau_p = 10$ and τ_p is increased by a factor of 10.

By what ratios have C_D and G changed?

5.5 Derive an equation for the low-frequency forward and reverse biased conductance of a p-n junction that includes generation and recombination in the depletion region.

6 / Switching Response

The p-n junction has many applications in which the device is used as an electrical switch. Typically a pulse of current or voltage is used to switch the diode from reverse bias, called the "off" state, to forward bias called the "on" state, and vice versa. Of prime concern to the circuit and device designer are the speeds at which the p-n junction can be made to switch states. We will first qualitatively discuss the transient which occurs when the diode is switched from the on to the off state and explain the origins of the time delay before the device turns off. The *reverse recovery time* is defined as the time delay from initiation of switching until the diode recovers to a specified degree of off, that is, a level of reverse current. The largest component of the reverse recovery time is the *storage time*. An approximate solution for the storage time is derived and compared to a more detailed mathematical solution. The transient response from reverse bias to forward bias is also discussed and an approximate solution is derived for one specific "turn-on" case.

6.1 THE TURN-OFF TRANSIENT

The dynamic switching of a diode from the forward conducting, or on state, to the off state is known as the turn-off transient. For the diode to be an *ideal* switch in the turn-off cycle, the current would traverse instantly from I_F, the forward dc value, to the reverse leakage current $-I_0$. It should be apparent from the discussions of Chapter 3, however, that the minority carrier charge stored in the bulk n- and p-regions must be removed before the device can be switched from forward conducting to reverse bias. Therefore, the real device cannot function as an ideal switch.

Figure 6.1(a) illustrates the idealized experiment for measuring the turn-off time of the diode and Fig. 6.1(b) is a sketch of the current response. The reverse current $-I_R$ is defined as the current an instant after the device is switched, while the *reverse recovery time* (t_{rr}) is defined as the time necessary for the diode current to recover to $-0.1\ I_R$. Figure 6.1(c) defines the *storage time* (t_s) as the time when the junction voltage $v_A(t)$ reaches zero volts or, as shown in Fig. 6.1(b), as the time associated with the nearly constant current ($-I_R$) of Fig. 6.1(b). The storage time, as will be discussed later, increases

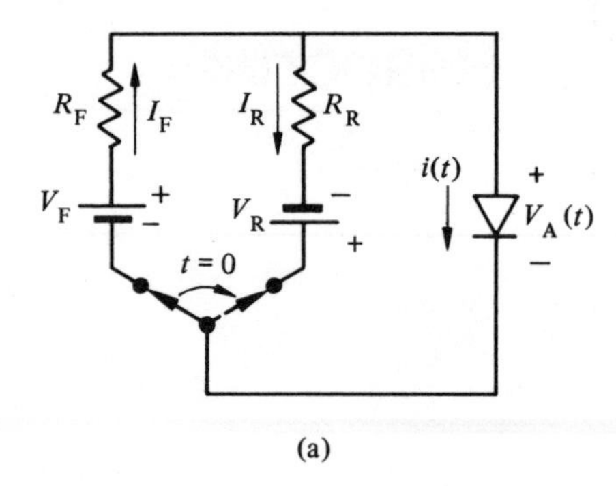

(a)

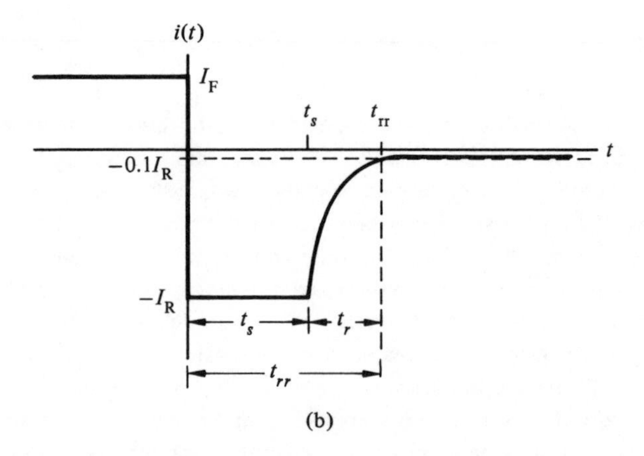

(b)

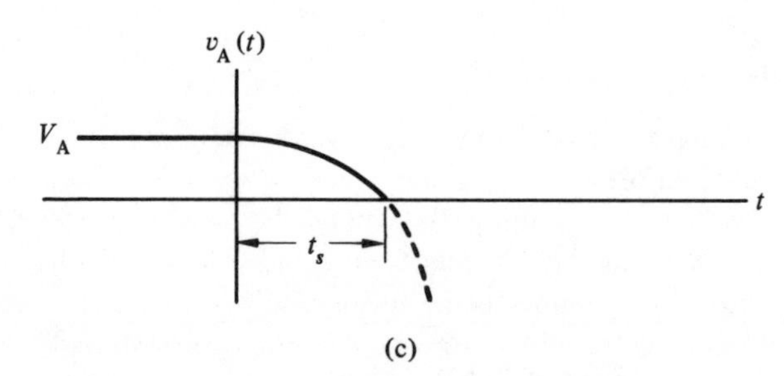

(c)

Fig. 6.1 Reverse recovery time and storage time: (a) circuit; (b) current; (c) voltage

with the amount of stored minority charge and is used as a figure of merit in switching applications. The *recovery time* (t_r) is the difference between t_{rr} and t_s.

The dc forward current (I_F) is obtained by inspection of Fig. 6.1(a) at $t = 0^-$, where V_F is assumed to be much greater than the dc diode voltage V_A. Since V_A is typically between 0.1 and 0.875 volts, the criterion is easily met in most cases where $V_F > 20$ volts. Writing a loop equation around the circuit, solving for the dc current, and assuming $V_A \ll V_F$ yields Eq. (6.1):

$$I_F = \frac{V_F - V_A}{R_F} \cong \frac{V_F}{R_F} \tag{6.1}$$

The reverse current (I_R) has a similar definition for $t = 0^+$, where the condition of $|v_A(t)| \ll V_R$ must be satisfied.

$$I_R = \frac{V_R + v_A}{R_R} \cong \frac{V_R}{R_R} \tag{6.2}$$

It should be noted that it is possible to make I_R much larger than I_F, a technique often used by switching-circuit designers to reduce the switching times.

The reader may initially be somewhat puzzled by Fig. 6.1(b) in which a large reverse current momentarily flows through the diode. Let's consider the physical explanation of the phenomenon before treating it analytically. The minority carrier concentrations before the device is switched are shown in Fig. 6.2. The total excess minority carrier charge stored in the junction is the area under each curve above p_{n0} and n_{p0}, respectively. Long after switching, at $t \rightarrow t_{rr}$, the minority carrier distributions are less than their thermal equilibrium values. Obviously the charge transfer necessary to turn the device off is represented by the cross-hatched areas. Remember that the minority carrier concentrations change in the quasi-neutral bulk regions by diffusing and/or recombination. Therefore to change the minority carrier concentrations of Fig. 6.2 from their $t = 0^-$ to $t \rightarrow t_{rr}$ values, current flow and recombination are necessary. Current flow, $-I_R$, is a result of holes leaving the n-region and electrons leaving the p-region by diffusion. Recombination occurs wherever the excess carrier concentrations exceed their thermal equilibrium values.

It should be apparent that the larger the minority carrier excess charge to be removed from the bulk regions, the longer the turn-off transient.

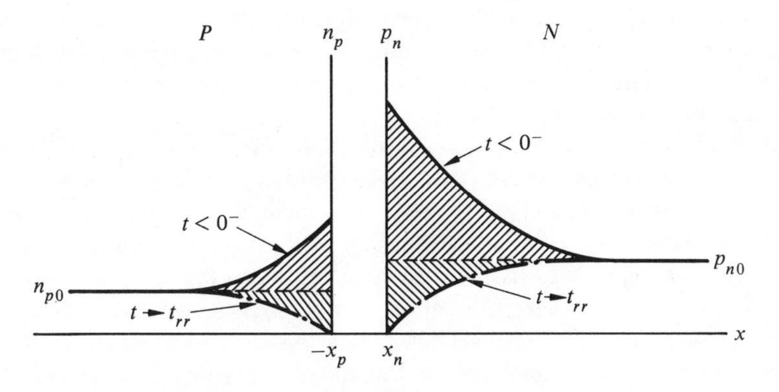

Fig. 6.2 Minority carrier charge storage.

Figure 6.2 can be used to deduce several qualitative relationships between the turn-off transient and the material properties. First if τ_p and τ_n are decreased (L_N and L_P are reduced), then less minority charge is stored at $t = 0^-$, and the turn-off time is made shorter because less charge transfer is necessary. Secondly, shorter lifetimes mean larger recombination of excess carriers, and hence a shortened turn-off time due to quicker charge removal. The circuit variables for shorter turn-off times are also closely coupled with the stored charge. A larger I_F produces a larger stored charge at $t = 0^-$ because a larger V_A is required and there is therefore a larger minority carrier concentration.

Now let's consider how the minority carriers change with time and distance as the device is switched from on to off. The derivation for dc currents in an ideal diode assumed that no generation or recombination occurred in the depletion region W. With this assumption the terminal current was obtained by adding Eq. (3.14) to Eq. (3.15), repeated here for the reader's convenience as Eq. (6.3).

$$J = -qD_P \frac{dp_n}{dx}\bigg|_{x_n} + qD_N \frac{dn_p}{dx}\bigg|_{-x_p} \tag{6.3}$$

Equation (6.3) is interpreted as stating that the slope of the minority carrier concentration, evaluated at the depletion region edge, determines the value of each current component. Also remember that the minority carrier concentrations at the depletion region edges were determined from Eqs. (3.30) and (3.28), repeated as Eq. (6.4) and (6.5) with V_A replaced by $v_A(t)$.

$$p_n(x_n) = p_{n0}e^{qv_A(t)/kT} = p_n(x_n, t) \tag{6.4}$$

$$n_p(-x_p) = n_{p0}e^{qv_A(t)/kT} = n_p(-x_p, t) \tag{6.5}$$

With the above equations in mind we are now in a position to explain the general nature of the switching response of the experiment shown in Fig. 6.1.

Figure 6.3 illustrates the time progression of excess charge removal with I_R at a constant value, that is, with a constant slope of the carrier concentrations at the depletion region edges. The value of I_R is nearly a constant as long as $|v_A(t)| \ll V_R$.

For the excess holes in the n-region, the constant diffusion current at x_n removes enough carriers to cause $p_n(x_n, t)$ to decrease in magnitude and there is hence a reduction in the terminal voltage according to Eq. (6.4). This reduction of $v_A(t)$ is illustrated in Fig. 6.1(c). Similar arguments for $n_p(-x_p, t)$ reiterate the reduction in excess carriers and the terminal voltage.

The storage time (t_s) is defined as occurring at the end of the constant current phase of the reverse current transient. Physically, the constant current phase ends when $|v_A(t)|$ becomes comparable to V_R and the two voltages subtract from each other, forcing $i(t)$ to be less negative than $-I_R$, as is illustrated in Fig. 6.1(b). Figure 6.3 pictures t_s as the time at which the junction voltage has become zero; that is, $p_n(x_n) = p_{n0}$ at $t = t_s$.

The remainder of the reverse current transient (t_r) is characterized by $v_A(t)$ becoming large and negative; that is, a reverse voltage develops across the junction. The current $i(t)$ is no longer a constant and is decreasing rapidly towards the steady state, reverse

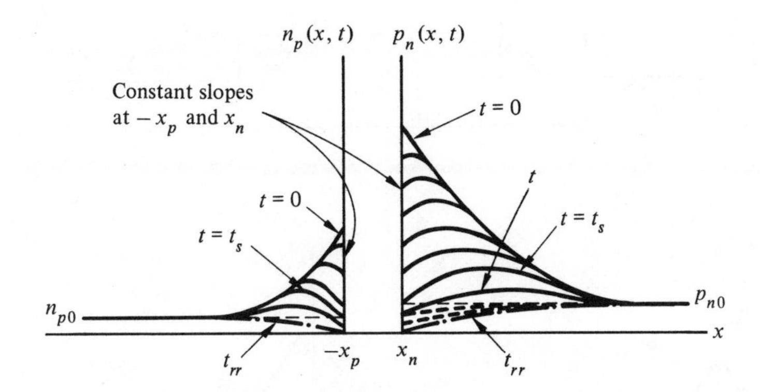

Fig. 6.3 Turn-off transient carrier concentrations.

leakage current $-I_0$. During this phase of the recovery the excess charge is being removed primarily by recombination (although some diffusion persists). Ultimately, the total excess charge is eliminated and the reverse bias causes the carrier deficit shown in Fig. 6.3 at $t = t_{rr}$. By the end of the reverse transient $v_A \cong -V_R$ and the device approaches the steady-state dc reverse bias condition.

The previous discussions allow us to predict some general results. The larger the forward current I_F the larger the minority carrier charge storage. Therefore, with a fixed I_R this means a larger t_s. A larger value of τ_p and τ_n means larger L_P and L_N values, resulting in more stored charge and again a larger t_s. In some fast-switching silicon diodes, the device is "gold doped" in addition to the n- and p-type impurities. Gold acts as a very effective recombination center near E_i in the band gap of silicon and thereby reduces the minority carrier lifetimes τ_n and τ_p, which in turn reduces t_s. Generalizing then, the larger the excess minority charge storage, the longer it takes to discharge the junction.

Finally, the storage time is also reduced by increasing the value of the reverse current I_R, pulling more charge per second from the bulk regions.

6.2 STORAGE TIME ANALYSIS

An approximate solution for the storage time can be obtained for the ideal step junction diode confirming many of our previous predictions. For simplicity, assume a p^+-n junction where the stored minority carrier charge is dominated by the holes in the bulk n-region. For this type device, as discussed in Chapter 3, the total current is approximately the hole current evaluated at x_n. Figure 6.4 illustrates the hole concentration, initially, and for the phases of the reverse recovery of the p^+-n diode.

The fundamental problem requires a solution for $\Delta p_n(x, t)$, as sketched in Fig. 6.4, a difficult problem indeed. We sidestep the problem by considering the total excess minority carrier charge $Q_P(t)$, remembering that the charge can change only through current flow $i(t)$ and recombination.

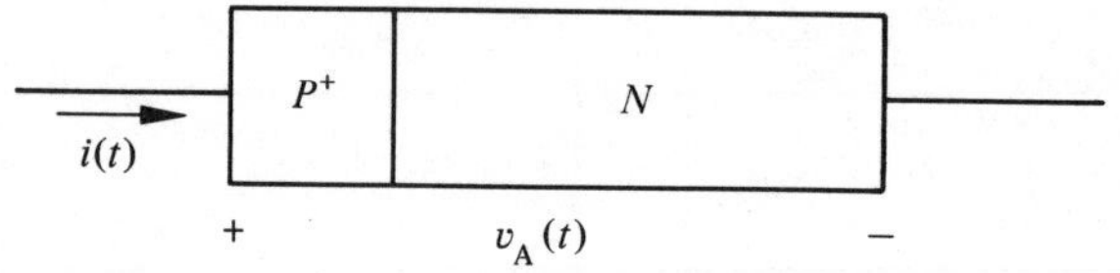

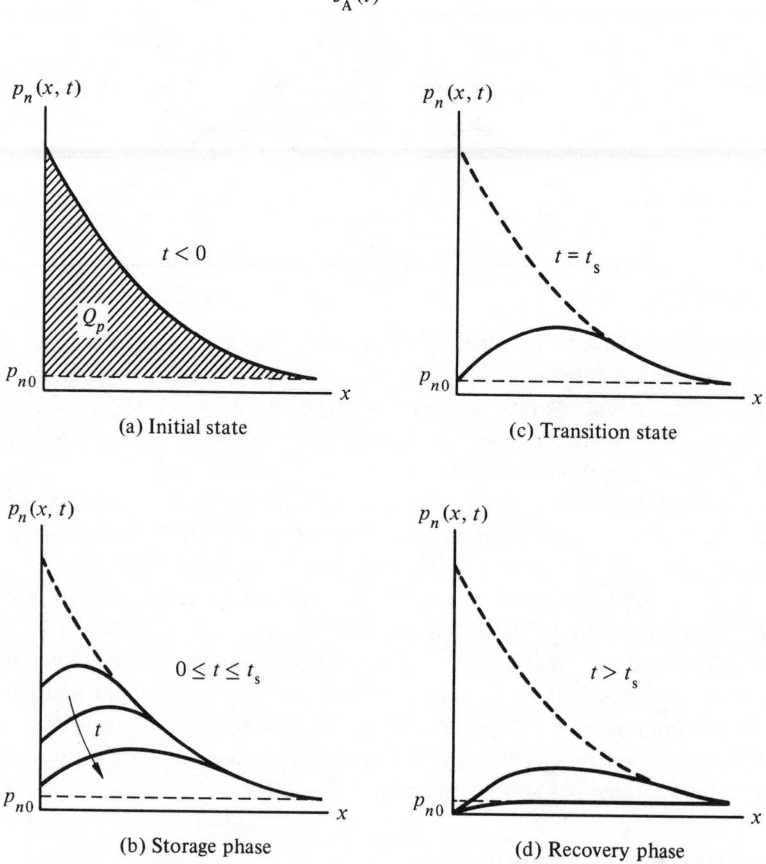

Fig. 6.4 Reverse recovery hole concentration profiles: (a) initial; (b) storage phase; (c) $t = t_s$; (d) recovery phase.

Volume I, Chapter 3, considered the *minority* carrier continuity equations for the bulk regions, repeated here for the n-bulk region, assuming low-level injection.

$$\frac{\partial \Delta p_n}{\partial t} = -\frac{1}{q}\frac{\partial J_P}{\partial x} - \frac{\Delta p_n}{\tau_p} \tag{6.6}$$

Equation (6.6) is multiplied by the area (A) and by q, then integrated over the n-bulk region

$$\frac{d}{dt}\left[qA\int_{x_n}^{\infty}\Delta p_n(x,t)\,dx\right] = -A\int_{J_P(x_n)}^{J_P(\infty)} dJ_P - \frac{1}{\tau_p}\left[(qA)\int_{x_n}^{\infty}\Delta p_n(x,t)\,dx\right] \tag{6.7}$$

It must be noted that the total excess hole charge is Eq. (6.8)

$$Q_P(t) = qA\int_{x_n}^{\infty}\Delta p_n(x,t)\,dx \tag{6.8}$$

Equation (6.7) can be rewritten as Eq. (6.9), after recognizing the $Q_P(t)$ terms of Eq. (6.8) as they appear in Eq. (6.7)

$$\frac{d}{dt}[Q_P(t)] = -A[J_P(\infty) - J_P(x_n)] - \frac{Q_P(t)}{\tau_p} \tag{6.9}$$

Remember that for the "long-base diode" the n-region is infinite in length; therefore $J_P(\infty)$ is zero since $d\Delta p_n/dx|_\infty = 0$. For the p^+-n junction the total current $i(t)$ is approximately $J_P(x_n)$; hence Eq. (6.9) becomes Eq. (6.10).

$$\boxed{\frac{dQ_P(t)}{dt} = i(t) - \frac{Q_P(t)}{\tau_p}} \tag{6.10}$$

Examination of Eq. (6.10) reveals the rate of change of charge storage is equal to the current (adding or subtracting charges per second) minus that lost to recombination. The total stored-hole charge changes by addition or removal of holes due to current and by recombination.

Let's now apply Eq. (6.10) to the approximate solution for the storage time of a p^+-n junction, using the circuit constraints of Fig. 6.1(a). For $t \geq 0$, but less than t_s the current $i(t) = -I_R$ in Eq. (6.10), as written in Eq. (6.11).

$$\frac{dQ_p(t)}{dt} = -I_R - \frac{Q_p(t)}{\tau_p}, \qquad 0^+ \leq t \leq t_s \tag{6.11}$$

Equation (6.11) can be solved directly as a differential equation or by applying a Laplace transform. It is a variables-separable type of differential equation. Separating the variables by multiplying by dt, then dividing by $I_R + Q_p(t)/\tau_p$ and integrating, we obtain Eq. (6.12a), where $Q_p\ (t_s)$ is assumed to be zero,

$$\int_{Q_p(0^+)}^{0}\frac{dQ_p}{\left[I_R + \dfrac{Q_p(t)}{\tau_p}\right]} = -\int_0^{t_s} dt = -t\Big|_0^{t_s} = -t_s \tag{6.12a}$$

Figure 6.4(c) shows that $Q_p(t_s) \neq 0$; however, to assume that it is zero results in a conservative estimate of t_s (a larger value). The term $Q_p(0^+)$ is illustrated in Fig. 6.4(a) as the initial hole charge stored in the bulk n-region. The left side of Eq. (6.12a) can be integrated from a standard table of integrals to be

$$\tau_p \ln\left(I_R + \frac{Q_p(t)}{\tau_p}\right)\Bigg|_{Q_p(0^+)}^{0} = -t_s = \tau_p \ln\left[I_R + \frac{0}{\tau_p}\right] - \tau_p \ln\left[I_R + \frac{Q_p(0^+)}{\tau_p}\right] \tag{6.12b}$$

Solving for t_s yields Eq. (6.12c),

$$t_s = -\tau_p \ln I_R + \tau_p \ln\left[I_R + \frac{Q_p(0^+)}{\tau_p}\right] \tag{6.12c}$$

therefore

$$t_s = \tau_p \ln\left[1 + \frac{Q_p(0^+)}{\tau_p I_R}\right] \tag{6.13}$$

Although $Q_p(t_s) \neq 0$ and, compared to $Q_p(0^+)$, it may not be negligible, the approximation nevertheless yields reasonable results; that is, the functional dependence is quite good.

For $t \leq 0$ Eq. (6.10) becomes Eq. (6.14), since the diode is at dc steady state and the stored charge is not changing with time,

$$\frac{dQ_p}{dt} = 0 = I_F - \frac{Q_p(0)}{\tau_p} \tag{6.14a}$$

Because the charge cannot change instantaneously, $Q_p(0^-) = Q_p(0^+)$ and solving for $Q_p(0^+)$ from Eq. (6.14a) yields

$$Q_p(0^-) = Q_p(0^+) = I_F \tau_p \tag{6.14b}$$

Substituting $Q_p(0^+)$ from Eq. (6.14b) into Eq. (6.13) yields Eq. (6.15).

$$t_s = \tau_p \ln\left[1 + \frac{I_F}{I_R}\right] \tag{6.15}$$

An examination of Eq. (6.15) shows that indeed our qualitative arguments are valid. In particular t_s will decrease if τ_p and I_F are decreased, which is the same effect as decreasing $Q_p(0^+)$. Note that t_s decreases if I_R is increased; this is identical to increasing the diffusion current to remove holes from the n-region faster.

A more detailed analysis of the charge storage in the p^+-n junction results in Eq. (6.16).

$$\text{erf}\sqrt{\frac{t_s}{\tau_p}} = \frac{1}{1 + \dfrac{I_R}{I_F}} \tag{6.16}$$

Figure 6.5 illustrates the difference between Eqs. (6.16) and (6.15), showing the conservative estimate for t_s obtained by Eq. (6.15). It should also be noted that the experimental measurement of t_s is used to obtain values of τ_p (or τ_n in the case of a p-n^+ device).

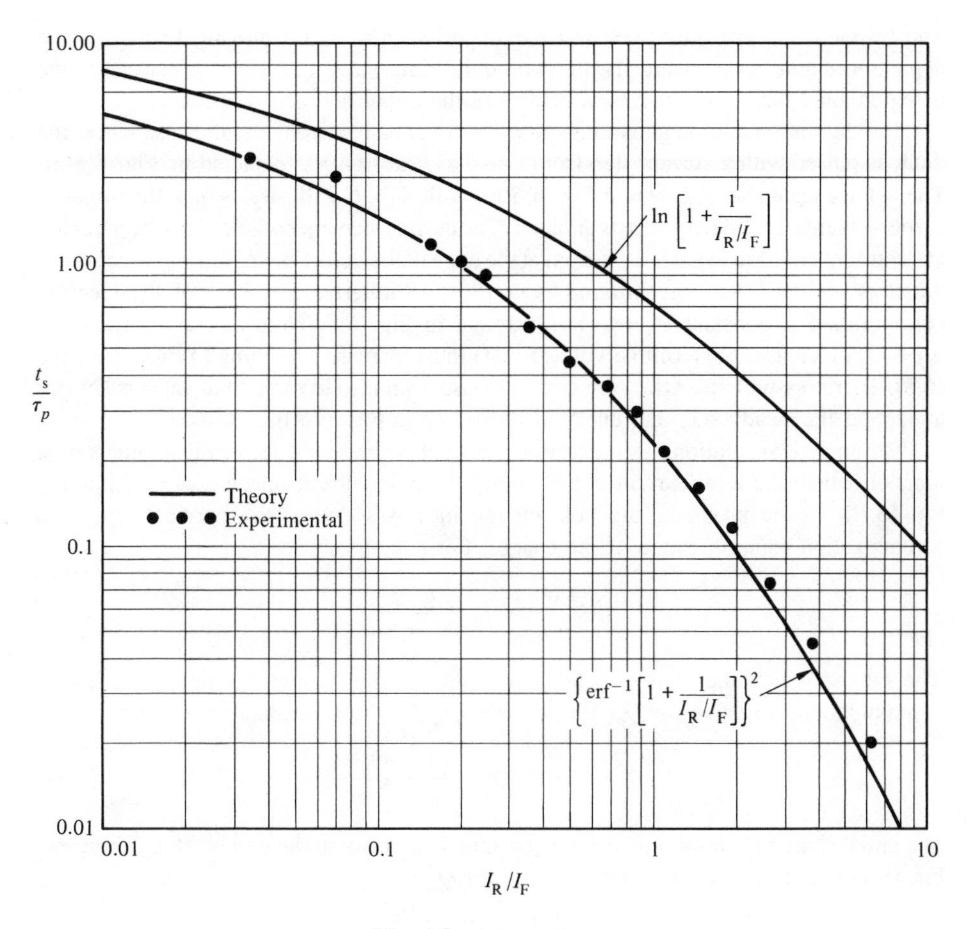

Fig. 6.5 t_s dependence

6.3 THE TURN-ON TRANSIENT

The forward turn-on transient occurs when the diode is pulsed from reverse bias into a forward current conducting state. This can be accomplished with a current pulse, a voltage pulse, or a mixture of the two pulses. Because of its simplicity and direct connection to many practical circuits, we will present here only the current pulsed case.

When the diode current is pulsed instantaneously from $-I_0$ to some constant forward current I_F, the voltage response, $v_A(t)$, changes from some negative value at $t = 0$ to V_A at $t = \infty$. The first stage of the response, from $t = 0^+$ to where $v_A = 0$, occurs very rapidly, in about the dielectric relaxation time of the semiconductor ($\cong 10^{-10}$ sec or less).

The response is very quick because the majority carriers are moving to narrow the depletion region back toward its thermal equilibrium value ($V_A = 0$). Electrons in the n-region and holes in the p-region neutralize the donor and acceptor ions.

Having disposed of negative bias to $v_A = 0$, let us next consider the case where the diode is pulsed with a current step from $i = 0$ to I_F at $t = 0$ as depicted by Fig. 6.6(a). The voltage response is shown in Fig. 6.6(b) while Fig. 6.7 illustrates how the minority carriers change until steady state is attained. The current step injects holes into the n-region at a constant rate. Since the hole current at the edge of the depletion region (x_n) is constant, the slope of the hole concentration is constant. Similarly, the slope of the electron concentration is constant at $-x_p$ as illustrated in Fig. 6.7. Note that as $p_n(x_n, t)$ and $n_p(-x_p, t)$ increase, the terminal voltage $v_A(t)$ must increase according to Eqs. (6.4) and (6.5). Therefore we expect the voltage to increase from zero to some final value necessary to support the steady-state current I_F as illustrated in Fig. 6.6(b).

The analytical solution for the turn-on transient is simplier if we again assume a p^+-n junction where the total current is essentially the hole current injected at x_n. Applying Eq. (6.10) to the problem, the hole charge increases due to the current step, with recombination using up some of the charge. For $t \geq 0$

$$\frac{dQ_p(t)}{dt} = I_F - \frac{Q_p(t)}{\tau_p}$$

The rate at which the charge builds up is due to the current I_F minus that lost to recombination. To solve for $Q_p(t)$ let's use a Laplace transformation.

$$sQ_p(s) - Q_p(0) = \frac{I_F}{s} - \frac{Q_p(s)}{\tau_p} \tag{6.17}$$

The initial charge $Q_p(0)$ is zero since no current flows prior to the current step. Therefore Eq. (6.17) becomes Eq. (6.19) by solving for $Q_p(s)$.

$$Q_p(s)\left[s + \frac{1}{\tau_p}\right] = \frac{I_F}{s} \tag{6.18}$$

$$Q_p(s) = \frac{I_F}{s(s + 1/\tau_p)} \tag{6.19}$$

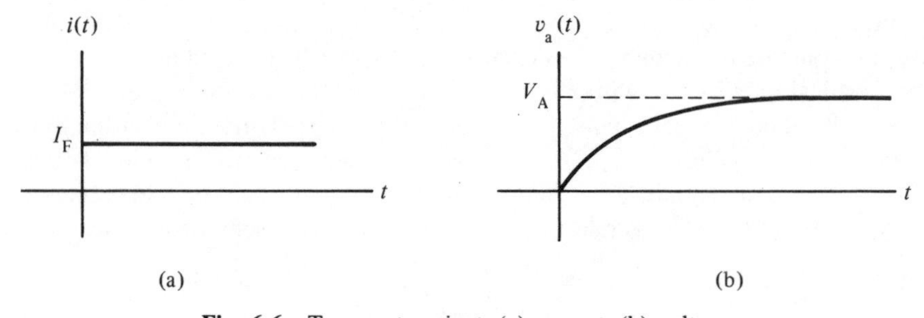

Fig. 6.6 Turn-on transient: (a) current; (b) voltage.

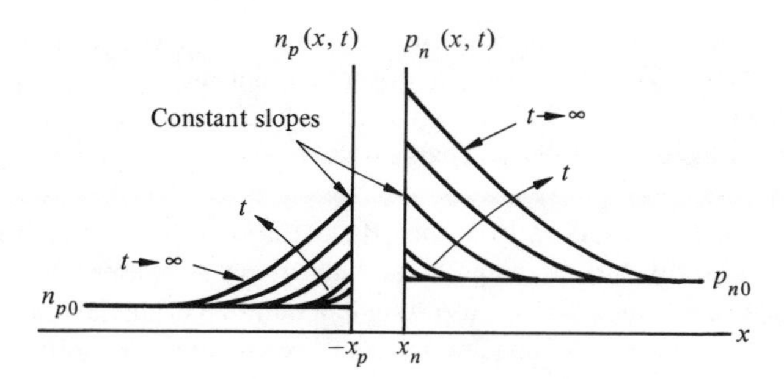

Fig. 6.7 Carrier concentrations during the turn-on transient.

The inverse Laplace transform of Eq. (6.19) yields $Q_p(t)$ as Eq. (6.20).

$$Q_p(t) = \tau_p I_F (1 - e^{-t/\tau_p}) \tag{6.20}$$

that is,

$$Q_p(\infty) = \tau_p I_F$$

from Eq. (6.20). Obviously the hole charge increases as the junction charges up to some final value.

To obtain an estimate of the voltage response shown in Fig. 6.6(b), let's assume that the hole distribution at each instant of time in Fig. 6.7 can be approximated by an exponential of the type

$$p_n(x', t) = p_{n0} e^{qv_A(t)/kT} e^{-x'/L_p} \tag{6.21a}$$

The total excess-hole charge is obtained by multiplying Eq. (6.21a) by qA and integrating over the entire bulk n-region, yielding Eq. (6.21b) as an approximation to $Q_p(t)$.

$$Q_p(t) = \tau_p I_F (1 - e^{-t/\tau_p}) \cong qA\, p_{n0} L_p (e^{qv_A(t)/kT} - 1) \tag{6.21b}$$

The voltage response can then be solved from Eq. (6.21b) as

$$v_A(t) \cong \frac{kT}{q} \ln\left[1 + \frac{\tau_p I_F}{qAp_{n0}L_p}(1 - e^{-t/\tau_p})\right] \tag{6.22}$$

Several observations can be made about the turn-on transient by inspection of Eq. (6.22). A smaller value of I_F and/or a smaller value of τ_p will give a faster turn-on time. Because $Q_p(\infty) = I_F \tau_p$, these results can be generalized as demonstrating that the smaller the amount of final stored charge, the quicker the device turns on. Again the amount of charge transfer controls the switching speed.

6.4 SUMMARY

The large signal pulsed response of the p-n junction was considered in terms of the turn-off and turn-on transient responses. When switched off from a conducting state, the current response was characterized by the reverse recovery time, t_{rr}. The longest component of t_{rr} is the storage time t_s. A reduction in the forward current and/or minority carrier lifetime reduces t_s. If the magnitude of the reverse current is made larger, t_s is made smaller. Any means of reducing the minority carrier charge storage before switching reduces the reverse recovery time.

The diode when switched on from the off state is controlled by the time necessary to build up the minority carrier charge in the bulk n and p regions. We considered the case of a current step from zero to I_F resulting in a build up of the junction voltage from zero to a final value of V_A. Again, any means of reducing the minority carrier charge storage will reduce the switching time. In particular, if I_F and/or τ_p and τ_n are reduced, the device will respond more quickly.

The original interest in diode transients was motivated by the desire to minimize the reverse recovery time and thereby reduce the switching time of diode logic circuits. Later, the reverse recovery transient found useful applications. Of particular importance is the step-recovery diode, which is fabricated in such a manner as to maximize the storage phase, t_s, and minimize the recovery phase, t_r. The charge storage characteristics are used to provide time delay elements in switching circuits and bidirectional current steering elements in diode access memories. The abrupt turn-off characteristic (with t_r usually less than 1 nsec) of the step recovery diode is used in fast, pulse shaping circuits and for efficient harmonic generation at high powers and frequencies.

PROBLEMS

6.1 If a p^+-n abrupt junction is conducting a current I_F in the forward direction and is then switched off through a current source, at $t = 0$, of

$$i_s = I_F e^{-t/\tau_p}$$

derive an equation for $Q_p(t)$ and give a rough sketch of the result.

6.2 A p^+-n junction is forward biased at I_{F_1} and pulsed to I_{F_2} at $t = 0$.

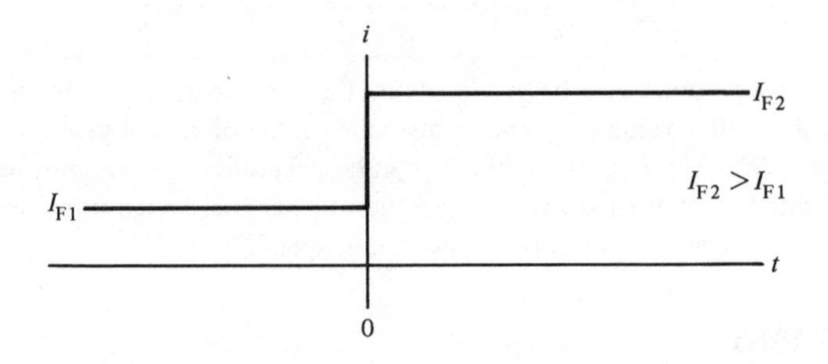

Figure P6.2

(a) Sketch the hole-carrier concentration as a function of time.

(b) Calculate $V_A(\infty) - V_A(0^-)$.

(c) Derive a formula for $Q_p(t)$.

6.3 A p^+-n step junction is switched from $I = 0$ to 1 mA with a current pulse. Calculate the time necessary for the device to reach 90% of its final voltage. $\tau_p = 1\ \mu$ sec, $I_0 = 10^{-15}$ A.

6.4 A p^+-n step junction is switched from a current source.

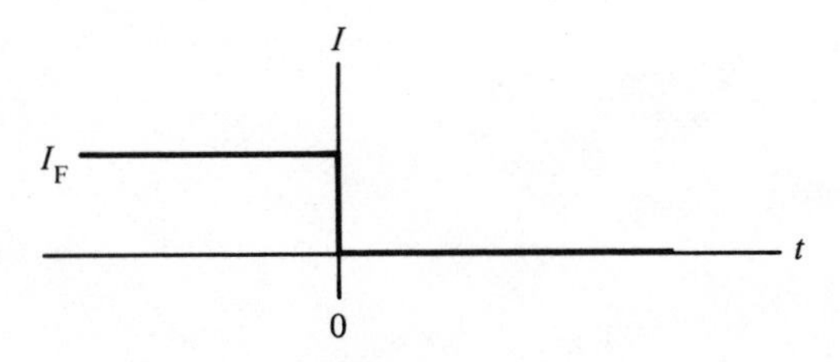

Figure P6.4

Derive an equation for $Q_p(t)$ and sketch $p_n(x, t)$.

Suggested Readings

1. P. E. Gray and C. L. Searle, *Electronic Principles*. New York: Wiley, 1969. Chapters 3 and 4 are an elementary presentation of the *p-n* junction.
2. W. H. Hayt and G. W. Neudeck, *Electronic Circuit Analysis and Design*. Boston: Houghton Mifflin, 1976. Chapters 1 and 2 are a good review of circuit models and diode circuits.
3. J. L. Moll, *Physics of Semiconductors*. New York: McGraw–Hill, 1964. Chapter 7 contains much more detailed discussions of generation and recombination in the depletion region, quasi-Fermi levels, and graded junctions.
4. B. G. Streetman, *Solid State Electronic Devices*. 2nd ed. Englewood Cliffs, N. J.: Prentice-Hall, 1980. Chapter 5 is a complete discussion of the diode from fabrication to charge storage. Chapter 6 presents the many applications of the *p-n* junction as light emitting, tunnel, solar, photo, and varactor diodes.

Volume Review Problem Sets and Answers

To work the following problems often requires an integrated knowledge of the subjects presented in Chapters 1 through 6. These problems serve as a general review of the subject matter.

PROBLEM SET A

A.1 List two methods used for fabricating an abrupt junction and two methods for a graded junction.

A.2 The thermal diffusion of impurities has a fixed number of the impurities. How do you expect the impurities to distribute themselves into silicon? Where is the metallurgical junction?

A.3 Why are regions I, II, and III nonideal?

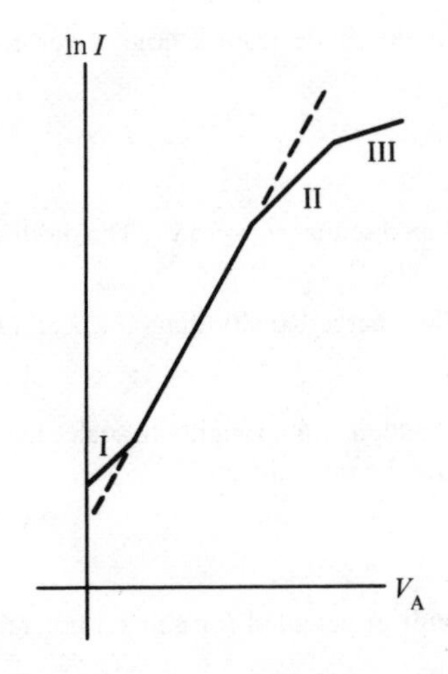

Figure PA.3

A.4 Explain the phenomena of "Zener" and "avalanche."

A.5 Sketch an energy band diagram for a p-n junction and explain what the "built-in potential" is and how it changes with increased doping of the acceptor material.

A.6 Explain *how* generation in the depletion region affects the $I-V_A$ characteristic in reverse bias.

A.7 For an abrupt p-n junction, sketch the minority carrier concentrations for forward bias, where $N_D > N_A$. Label *all* the boundary conditions and excess carrier concentrations, etc.

A.8 If N_D increases ($\uparrow$), how will the following parameters vary for a p^+-n junction?

τ_p $\quad$ τ_n $\quad$ C_J $\quad$ C_D $\quad$ $\mathscr{E}(0)$

A.9 A p-n junction is such that $L_P \gg W_N$ and $L_N \gg W_P$, and

$$\Delta p_n(W_N) = 0$$
$$\Delta n_p(-W_P) = 0$$

Sketch the minority carrier distributions for forward and reverse bias.

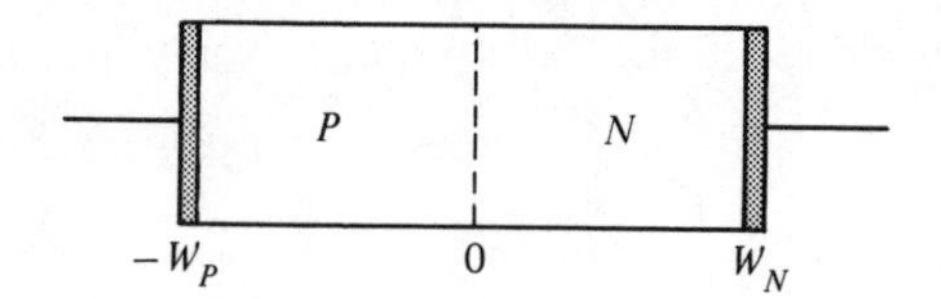

Figure PA.9

A.10 A p-n^+ junction is forward biased. Sketch the junction's conductance and diffusion capacitance vs $\omega\tau_n$.

A.11 A p-n^+ junction is forward biased at I_F when open circuited at $t = 0$. Sketch the minority carrier distribution $\Delta n_p(x, t)$. Also sketch $v_A(t)$.

A.12 Explain why switching of the diode from a large reverse bias to $V_A = 0$ is quite rapid.

PROBLEM SET B

B.1 The n-p junction diode is doped so that $N_D = 5N_A$. The diode is at thermal equilibrium. *Label* all significant points.

(a) Sketch (roughly to scale) the charge-density diagram. Let $|qN_A|$ equal one unit.

If $E_G = 4$ units on the graph:

(b) sketch the electron energy-band diagram roughly to scale. Let $|E_F - E_V|$ on the p side be one unit and $|E_c - E_F|$ on the n side be one-half unit.

(c) Sketch the electric field.

(d) Sketch the potential.

(e) What is the value of the built-in potential (or contact potential), in # of squares?

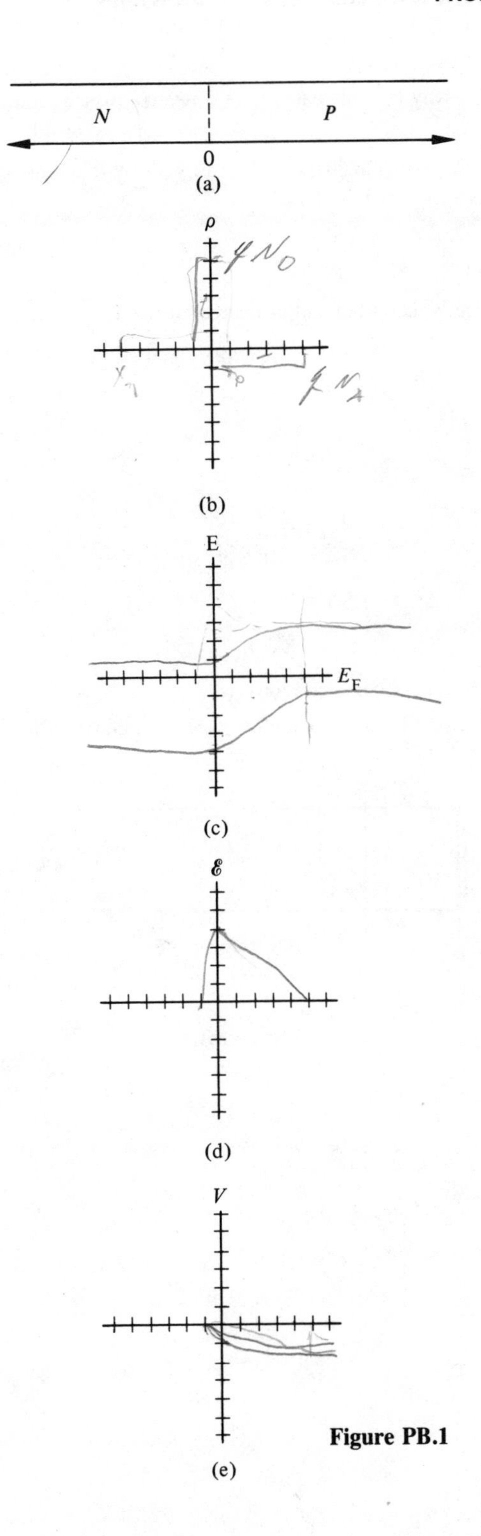

Figure PB.1

B.2 List several reasons why the calculated I_0 of a reverse biased diode does not agree with the measured value.

B.3 A p-n abrupt junction is symmetrically doped ($N_A = N_D$) and driven from a current source that is stepped from zero to I_F at $t = 0$.

(a) Sketch $Q_N(t)$ and $Q_P(t)$.

(b) Sketch $n_p(x, t)$ and $p_n(x, t)$.

(c) Explain how charge is increased and/or decreased from $t = 0$ to $t = \infty$.

B.4 If $Y = 10^{-3}\sqrt{1 + j\omega 10^{-7}}$ for a p^+-n junction, calculate G and C_D for $\omega = 10^7$ rad/sec and 5×10^7 rad/sec.

B.5 *Forward bias diode:*

$n_i^2 = 10^{20}/\text{cm}^3$, $\quad \tau_n = 10^{-7}$ sec,

$N_A = 10^{16}/\text{cm}^3$, $\quad \tau_p = 5 \times 10^{-7}$ sec,

$N_D = 10^{14}/\text{cm}^3$, $\quad D_P = 20\ \text{cm}^2/\text{sec}$,

$D_N = 50\ \text{cm}^2/\text{sec}$, $\quad \Delta n_p(-W_N) = 0$,

$W_N = 2 \times 10^{-4}$ cm

(a) Sketch the minority carrier concentrations.

(b) If no recombination occurs in the p-region, derive an equation for $\Delta n_p(x)$.

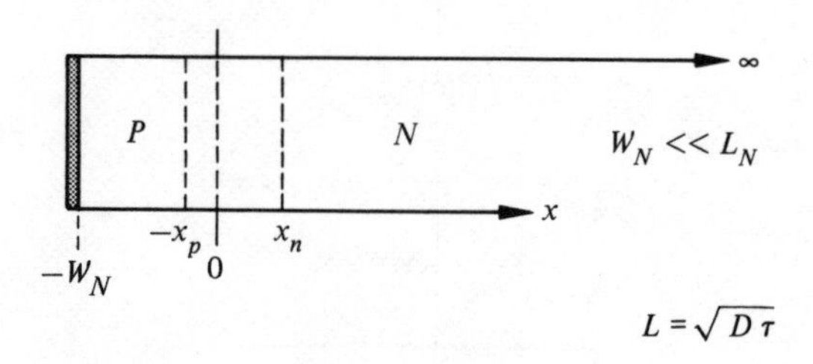

Figure PB.5

ANSWERS TO SET A REVIEW PROBLEMS

A.1 Alloy Diffusion
↕ ↕
Epitaxial Ion implant

A.2 Gaussian, $N(x) = N_B$.

A.3 I, recombination in W; II, High injection; III, R_S.

A.4 Zener is tunneling; avalanche is where the carrier ionizes a silicon atom by impact collision to free an electron–hole pair.

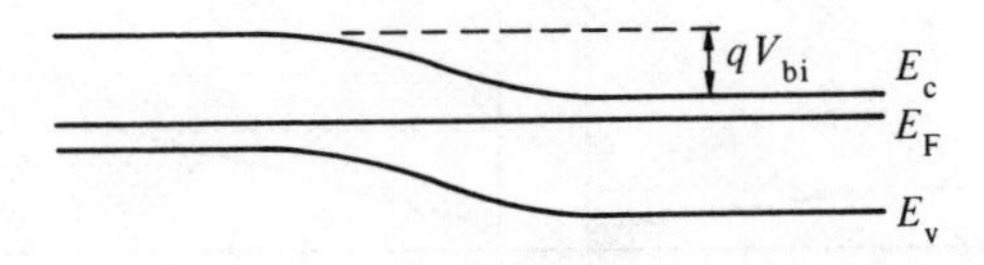

Figure AA.5

A.5 V_{bi} increases with N_A or N_D.

A.6 Adds carriers generated in W to the current flow; reverse current is increased and gets larger with V_A.

A.7

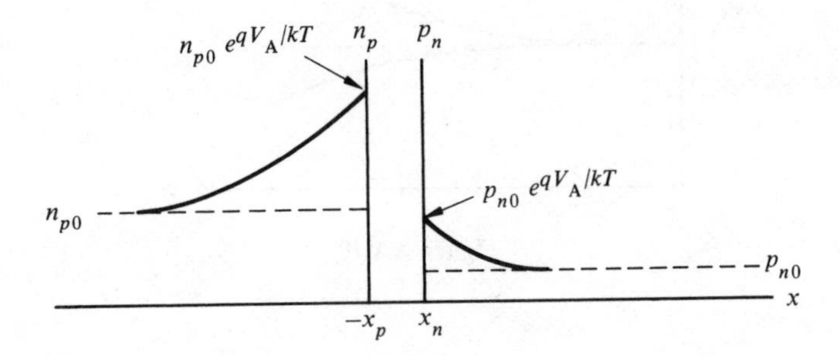

Figure AA.7

A.8 τ_p decreases, τ_n not effected, C_J and $\mathscr{E}(0)$ increase, C_D decreases.

A.9

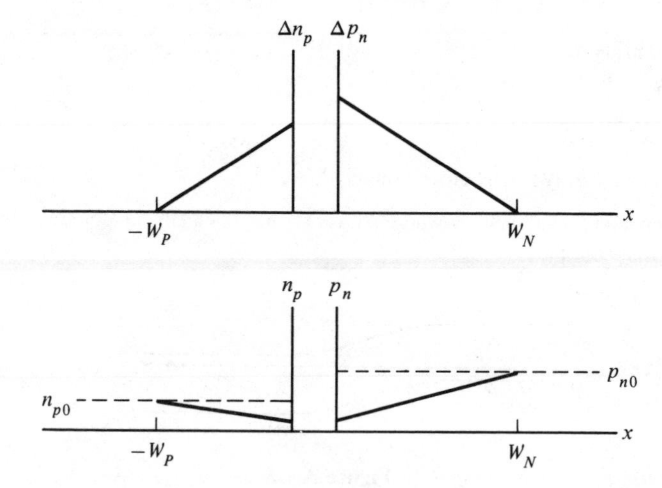

Figure AA.9

A.10

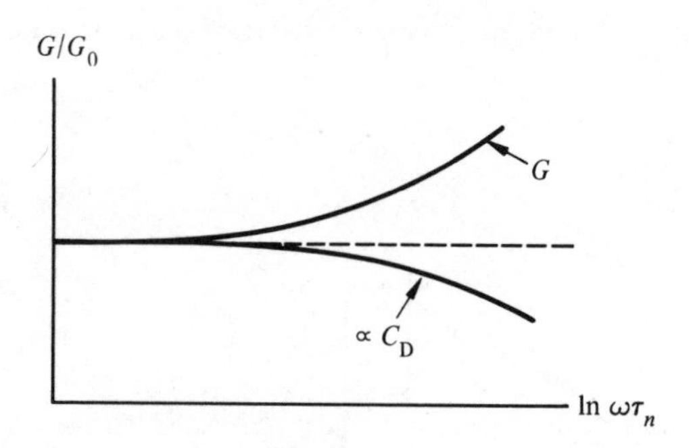

Figure AA.10

A.11

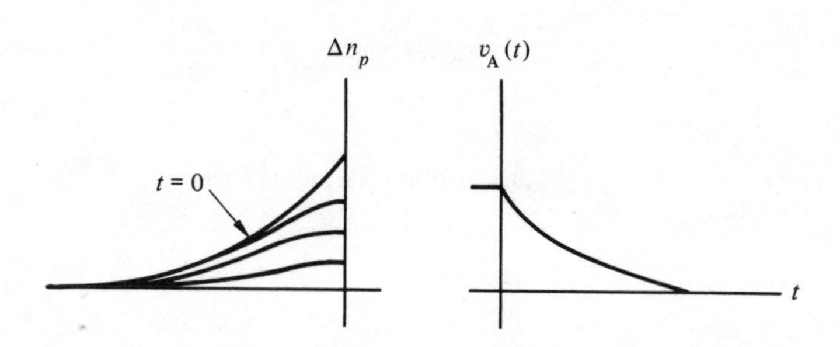

Figure AA.11

A.12 Because the depletion width change is the movement of majority carriers that change at about the dielectric relaxation time.

ANSWERS TO SET B REVIEW PROBLEMS

B.1

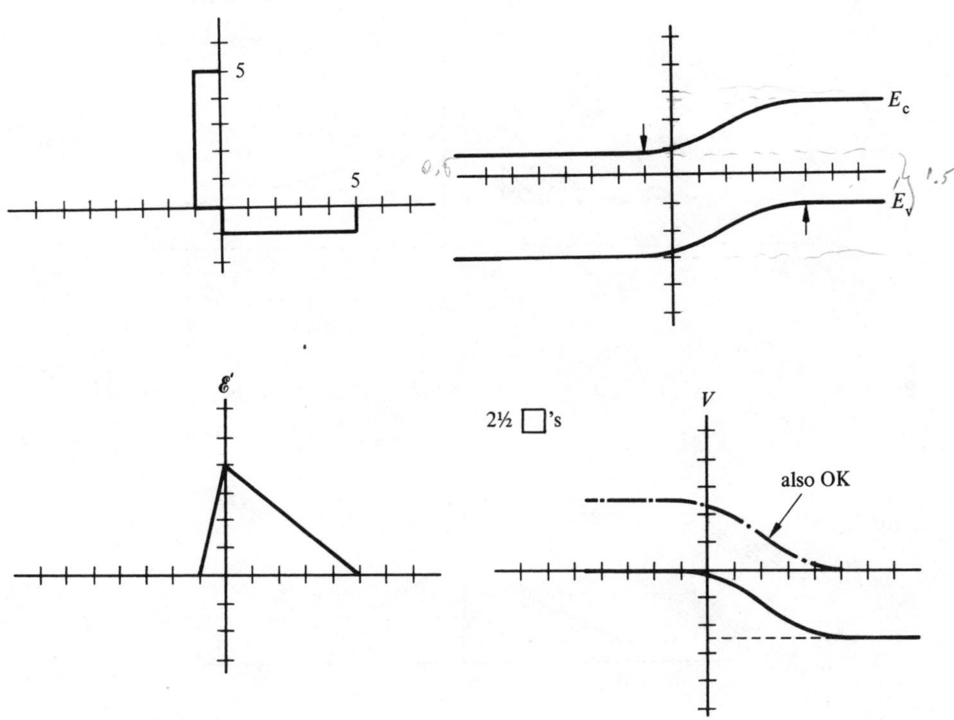

Figure AB.1

B.2 Generation in W, surface leakage

B.3 Slope at $x = 0$ is constant! I_F injects charge and builds up with time; recombination limits the build up and at t = infinity the charge entering = recombination.

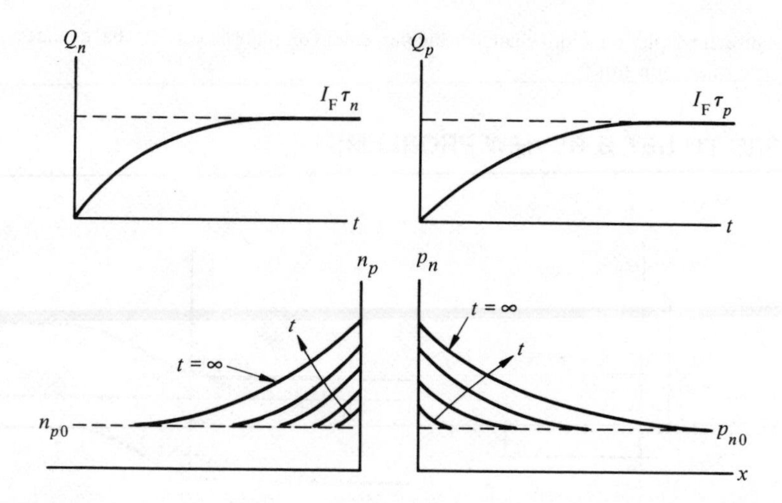

Figure AB.3

B.4 $G = 1.0986$ m mho, $C_D = 45.5$ pF; $G = 1.746$ m mhos. $C_D = 28.6$ pF.

B.5

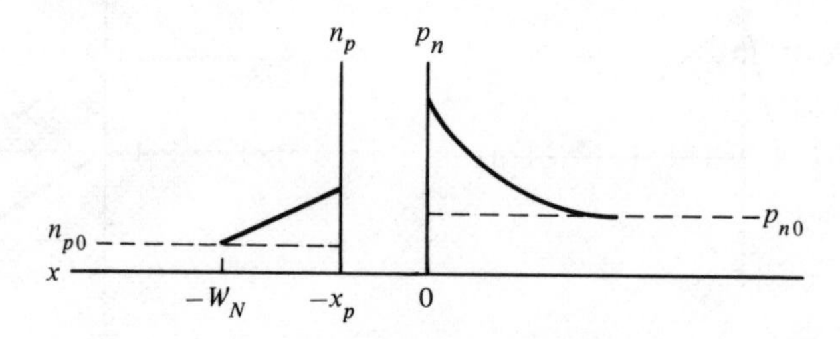

Figure AB.5

$$t_n \to \infty \therefore D_N \frac{d^2 \Delta n_p}{dx^2} = 0$$

$$\Delta n_p(x) = n_{p0}\,(e^{qV_A/kT} - 1)\frac{x + W_N}{W_N - x_p}$$

Appendix

LIST OF SYMBOLS

a	grading constant, linearly graded (#/cm^4)
A	cross-sectional area (cm^2)
C	capacitance (F)
C_J	junction (depletion) capacitance
C_{J0}	junction (depletion) capacitance at $V_A = 0$
C_D	diffusion capacitance
D	impurity diffusion constant (cm^2/sec)
D_N	electron diffusion constant (cm^2/sec)
D_P	hole diffusion constant (cm^2/sec)
E_C	conduction band edge energy level (eV)
E_F	Fermi energy level (eV)
E_i	intrinsic Fermi energy level (eV)
E_V	valence band edge energy level (eV)
E_{CN}	conduction band edge in n-material
E_{CP}	conduction band edge in p-material
E_{VN}	valence band edge in n-material
E_{VP}	valence band edge in p-material
G	generation rate for electrons and holes (#/cm^3-sec)
G	conductance (Ω)
G_0	low-frequency conductance
i	diode signal current

I	total current (A)
I_N	current due to electrons (A)
I_P	current due to holes (A)
I_0	diode reverse saturation (leakage) current (A)
I_{R-G}	recombination–generation current component (A)
J	total current density (A/cm^2)
J_N	electron current density (A/cm^2)
J_P	hole current density (A/cm^2)
k	Boltzmann's constant
K_O	relative dielectric constant of oxide
K_S	relative dielectric constant of semiconductor
kT	thermal energy (eV)
L_N	electron minority carrier diffusion length (cm)
L_P	hole minority carrier diffusion length
L_P^*	complex diffusion length for holes
M	multiplication factor
m	exponent in multiplication
m	depletion capacitance exponent, $1/3 \leqq m \leqq 1/2$
N_A	acceptor impurity concentration (#/cm^3)
N_D	donor impurity concentration
$N(x,t)$	impurity concentration (#/cm^3)
N_0	surface concentration (#/cm^3)
N_A^-	ionized acceptor concentration (#/cm^3)
N_D^+	ionized donor concentration
n	ideality factor for a diode
n_0	electron concentration at thermal equilibrium
n_n	electron concentration in n-material (#/cm^3)
n_p	electron concentration in p-material (#/cm^3)
n	electron concentration (#/cm^3)
n_{n0}	electron concentration in n-material at thermal equilibrium
n_{p0}	electron concentration in p-material at thermal equilibrium
n_i	intrinsic carrier concentration (#/cm^3)
n^+	degenerately doped n-type material
p^+	degenerately doped p-type material
p_p	hole concentration in p-material (#/cm^3)
p_n	hole concentration in n-material (#/cm^3)
p	hole concentration (#/cm^3)

p_{n0}	hole concentration in n-material at thermal equilibrium
p_{p0}	hole concentration in p-material at thermal equilibrium
p_0	hole concentration at thermal equilibrium (#/cm^3)
$\tilde{p}_N(x, t)$	signal component of hole concentration
$\hat{p}_N(x)$	x variation of hole signal component
Q	total number of impurities initially deposited at the surface
Q_p	hole minority carrier charge (coul)
q	charge on an electron (1.602×10^{-19} coul)
R_s	diode series resistance (bulk and contact) (Ω)
r	dynamic (small signal) resistance
t	time (sec)
T	temperature in degrees Kelvin (°K)
t_s	diode storage time
t_{rr}	reverse recovery time
t_r	recovery time
V_A	diode applied voltages (V)
V_{bi}	junction built-in voltage (V)
$V(x)$	potential (V)
V_N	contact potential of n-material to metal (V)
V_P	contact potential of p-material to metal (V)
V_j	junction voltage across depletion region (V)
V_{BR}	diode breakdown (avalanche or Zener) voltage
v_A	total instantaneous applied voltage
V_A	dc value of applied voltage
v_a	small signal applied voltage
W	depletion region total width (cm)
x_n	n-region depletion region width (cm)
x_p	p-region depletion region width (cm)
x_j	junction depth (cm)
x, x', x''	x-axis variable (cm)
Y	admittance ($1/\Omega$)
ρ	charge density (coul/cm^3)
ω	radian frequency (rad/sec)
∇	gradient symbol
$\mathscr{E}$	electric field (V/cm)
$\mathscr{E}_{CR}$	electric field at avalanche

ε_0	permittivity of free space 8.854×10^{-14} (farad/cm)
μ_n	electron mobility (cm^2/V-sec)
μ_p	hole mobility (cm^2/V-sec)
τ_n	electron minority carrier lifetime (sec)
τ_p	hole minority carrier lifetime/(sec)
τ_0	effective lifetime in depletion region
Δn_p	excess electron concentration in p-material
Δn_n	excess electron concentration in n-material
Δp_n	excess hole concentration in n-material
Δp_p	excess hole concentration in p-material
$\overline{\Delta p_n}(x)$	the dc value of excess hole concentration

Index